THE BAY

PUBLISHER'S NOTE

Works published as part of the Maryland Paperback Bookshelf are, we like to think, books that have stood the test of time. They are classics of a kind, so we reprint them today as they appeared when first published many years ago. While some social attitudes have changed and knowledge of our surroundings has increased, we believe that the value of these books as literature, as history, and as timeless perspectives on our region remains undiminished.

ALSO AVAILABLE IN THE SERIES:

The Amiable Baltimoreans by Francis F. Beirne
Tobacco Coast by Arthur Pierce Middleton

A Naturalist discovers a universe of life above and below the Chesapeake

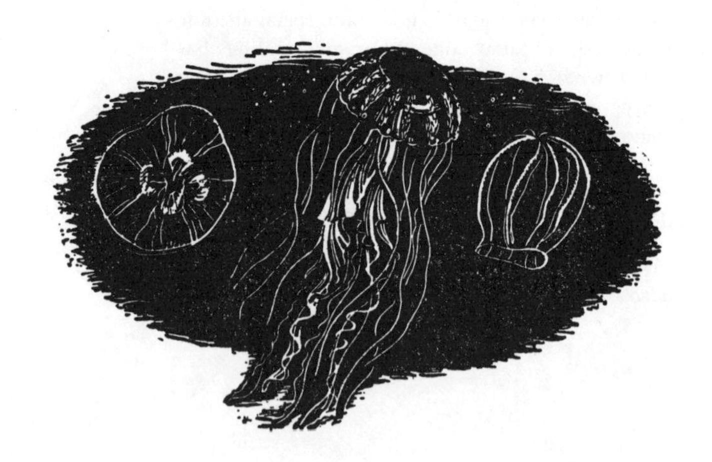

THE BAY

By Gilbert C. Klingel

ILLUSTRATED BY NATALIE HARLAN DAVIS

The Johns Hopkins University Press

BALTIMORE AND LONDON

To Emily, whose wise tolerance and understanding have made her the ideal sister, this volume is affectionately dedicated

Originally published in 1951 by Dodd, Mead & Company, New York

Maryland Paperback Bookshelf edition published in 1984 by
The Johns Hopkins University Press

Printed in the United States of America

The Johns Hopkins University Press, Baltimore, Maryland 21218
The Johns Hopkins Press Ltd., London

The paper in this book is acid-free and meets the guidelines for permanence and durability of the Committee on Production Guidelines for Book Longevity of the Council on Library Resources.

Library of Congress Cataloging in Publication Data

Klingel, Gilbert C.

The Bay: a naturalist discovers a universe of life above and below the Chesapeake.

(Maryland paperback bookshelf)

Reprint. Originally published: New York: Dodd, Mead, c1951.

1. Natural history—Chesapeake Bay (Md. and Va.)
2. Natural history—Outdoor books. I. Title. II. Series.
QH104.5.C45K5 1984 508.755'18 84-47954
ISBN 0-8018-2536-9 (pbk.: alk. paper)

PREFACE

THIS BOOK IS ABOUT A WONderful place. It is called, after the manner of the Indians who first peopled its shores, *The Chesapeake.* The modern natives, the tidewater Marylanders and Virginians who live by it and on it, refer to it more simply as *The Bay.*

A number of good books have been written about the Chesapeake Bay region, particularly about the somewhat fabulous colonials who established themselves in magnificent estates along its rivers and inlets in the seventeenth and eighteenth centuries. Similarly, the equally romantic sailing ships and seafaring men who were a product of the Bay have received their due share of attention. A number of technical and scientific treatises about various aspects of the geology and biology of the area have been published, but, like most technical works, they stand motionless on various library shelves. This volume deals with none of these and does not pretend to be scientific. Instead it is concerned with a subject which most of these earlier volumes have curiously neglected—the Bay itself.

That the Chesapeake as a natural phenomenon should have been slighted in our more readable literature is surprising, for the Bay is one of the most remarkable bodies of water in the world and the lives of many of the creatures that have their being in its depths are little short of marvelous.

I have wanted to write of this Bay because for the past twenty years I have contemplated its flow with increasing wonder and because the story of the Chesapeake should be told. In these pages I have attempted to set down on paper something of the character of the Bay and something of its moods as I have seen them in these past two decades—the sleek still calms, the gray fogs, the sparkling lusty northwesters, the goodly smell of the swamps, and the clean feel of the salt air. Beneath the dark-green waters strange events take place. Few have witnessed them, as most of this undersea activity is hidden and concealed. Only to those rare individuals who, with undersea equipment, have ventured beneath the surface has a portion of this flow of life been revealed. I have been privileged to spend many hours on the floor of the Chesapeake in varied types of diving equipment and it is partly the purpose of this book to record some of the sights witnessed there. While a number of books have been written in recent years about the world of the undersea most of these have been about the waters of the tropics. Almost no effort has been made to explore the depths of our own, American, waters. These in their own peculiar and somber way are as entrancing as those of any foreign land.

Much that has made the Chesapeake unique is changing and I hasten to record before the alterations are complete. Even now places that were solitudes are no longer lonely and beaches and rivers that were havens of peace have become otherwise. However, it is certain that, although the surrounding country may alter materially, the great body of the Chesapeake will remain much as it has been for centuries past and much as it will be when man and his works shall have been long forgotten. For the Bay is born of the ocean and the ocean does not change.

CONTENTS

CHAPTER 1

THE INCREDIBLE CHESAPEAKE

IN THE SHORT SPACE OF FIVE minutes the earth has entirely altered. The moonlight that a few moments ago was flooding the waters of the Chesapeake has disappeared and in its place is only a wan evanescent glow, intangible and barely to be distinguished. The stars that had been gleaming brightly have gone as if they had never existed. From the warm soft air of an early May night I have become enveloped in a chill that creeps into the flesh and causes a shudder to course over my body.

In a minute more even the glow is gone and I am surrounded by darkness, a blackness relieved only by pale ghostly lights which dim and then flare again for brief seconds in the hazy distance. Even these are vague and there is no direction or object upon which to fix time, space, or dimension. Although the brightly lighted, teeming streets of Baltimore City with its near million population are only sixty miles away and although Washington is less than half this distance I feel as far removed as though I have entered another world.

And, in a sense, I have entered another world, for I am swinging from the end of a long rope twenty feet beneath the surface of the Chesapeake Bay, about three miles out from shore and about five miles from the mouth of the Potomac River. In these five minutes I have come from a modern understandable sphere to one of strangeness and unreality.

Swaying gently back and forth as though on a pendulum, I peered into a great void. As I watched, something ghostly flitted across the glass of my bronze helmet, burst into livid green flame, and then as abruptly disappeared. Presently out of this black space came other green flashes, some near and some far; then there was created before my eyes a galaxy of tiny lights as though Lilliputian stars were bursting and disappearing into space.

I felt something slither across my bare arm and the object, whatever it was, moved over my shoulder and sped before my face, leaving a trail of green color. In a second this faded and once again I found myself shrouded in gloom. Tilting my body back I peered toward the surface. Very faintly a slight change in hue indicated that far above there still existed the world of the open air and the bright moonlight. I turned down again. Between my feet the rope coiled into the dark faintly marked by a long line of phosphorescence. Soon I became aware that I, too, was bathed in greenish light, by the glow of millions of minute animals excited by my presence.

It had seemed a simple thing in the early hours of that May evening to step into a small boat and sail out on the Chesapeake Bay, to anchor, don a diving suit, and drop beneath the waters. But now that the surface was twenty feet above my head and there lay thirty feet of black Bay below, I found that I was beginning to feel a little overawed, and this feeling was enhanced by the knowledge that, although many thousands of people in steamers and pleasure yachts had sailed over this very spot, I was the first human being ever to penetrate intimately this particular portion of the earth's surface.

I let the rope slide through my fingers and dropped slowly into the depths. At thirty feet the water became piercingly chill and the tide, which had been quiet and still near the surface, was flowing toward the open sea so strongly that it was with some difficulty I maintained my position on the lifeline. The water that a short time before had been only faintly lighted by phosphorescence was ablaze. In a few seconds I found myself shivering and gasping for breath within the confines of my helmet. Tilting my head upward to keep the helmet free of water, I floated the last ten feet downward and then felt my feet gently hit the yielding sand of the sea floor. Still struggling against the current I crouched to my knees, and then to keep from being carried away I flattened myself on the bottom. Digging my toes and fingers into the silt

I managed to seize something firm and with my free hand I groped in my belt for a watertight flashlight.

Instantly a long streamer of white pierced the dark. The beam accentuated the black masses of water on either side. The bottom was fine, hard, yellowish sand; the grains were shifting with the current and were rolling and bumping into each other. The sea floor was etched in deep furrows, long lines that wound in regular sequence across the path of the tide. In the rays of light these appeared as tall hills and deep valleys steeped in shadow. Turning the light around I found that the object I was grasping was a small purplish finger sponge which had anchored itself to some invisible material beneath the sand; to this, like myself, it was clinging to hold against the current. The finger sponge was a lovely object, a queer thing of red and purple looking like some Martian plant transposed to earth.

Not far distant was another of these vegetablelike sponges and in the lee of this sponge showed up a creature with a most droll appearance. Its face, if face it could be called, was covered with plates of mail; these were deeply etched with engraving which resembled the patterns the frost makes on the windowpanes on a wintry morning. The plates ended in sharp spines and odd-shaped protuberances. And set in this chased and armored face, regarding me steadily, was a pair of soft eyes which glowed greenishly.

Most interesting, however, was not the creature's face but the disproportion of its pectoral fins which stood out stiffly from the sides as though they were large wings. Still more startling was the fact that the fish, for fish it was, was not swimming as fishes normally do but was walking delicately on six long flexible rays which appeared to grow from the base of the head at the junction of the winglike pectoral fins.

Apparently intrigued by the novelty of my lamp the creature stepped daintily forward, first extending one ray, then the others, in a mincing gait which reminded one of an old-fashioned minuet. Fastidiously it inched its way forward to where my fingers lay extended in the sand. When it reached them it carefully strode up to each in turn, and then gently with its mouth nudged the tips. By way of experiment, I quickly raised my fingers to see what would happen. Instead of swimming away, as one might expect, the creature simply sprang to one side, rose about a foot in the water, and then on the widespread pectoral fins floated gently to the bottom. When it alighted the rays sent up a little cloud of sand particles which swirled away in the dark.

Once again the sea robin came close, walked over the back of my hand, tickling the flesh as it passed. Then without apparent cause it suddenly turned and streaked out of sight.

The ocean floor for a time seemed lifeless and I once again became conscious of the cold. It did not seem possible that only fifty feet above my head a soft warm air smelling of myrtle, pine trees, and with the scent of locust was stirring gently over the Bay.

Shivering, I focused my attention again on the sea bottom. Although the sand was drifting slightly and although the uppermost grains were rolling and bumping into one another, I could see that even in the short radius of my vision the sand bottom was peopled with a host of creatures. The creatures themselves were not visible but their presence was indicated by large numbers of holes and cavities in the sea floor. Some of these holes were no bigger than the lead of a pencil, but there were a few into which I could have thrust my thumb. Accentuated by the ray of the lamp and appearing much as specks of matter appear in a narrow field under a microscope, tiny particles were surging in and out of these holes as the inhabitants sucked in or ejected the water. With my fingers I tried turning up the bottom, hoping to cast loose its inhabitants, but all I got for my pains was a cloud of silt which rose like smoke, obscured my vision, and then slowly drifted away.

With this plowing of the bottom I must have disturbed some of these invisible beings, and no doubt the scent of their torn tissues must have gone sweeping along with the silt on the tide, because in a few seconds I was surrounded by darting forms. They were small fishes. Like so many dwellers in the undersea, which are pale and dull when viewed in the upper air, these fishes seen in their own element and lighted with the rays of the lamp were creatures of surprising beauty. Although their bodies were only an inch or so in length, and they seemed at times almost translucent, their scales flashed delicate iridescent pinks and lavenders tinged with overtones of shimmering blues and with glaucous greens.

Soon new forms of life began to appear. In front of my light little pinpoints of light darted, the flashing reflection of hordes of tiny crustaceans which had been attracted by the rays. The light appeared to excite them, for they streaked and tumbled as though in some queer kind of crustacean frenzy. Some shot headlong and flung themselves at the bulb; others began whirling in circles, going round and round as though activated by invisible springs. The number increased momentarily until the long cone of light became a mass of twisting organisms.

Among the larger were some glass prawns, shrimplike creatures, and, as their name suggests, they were as clear as filtered water. Through their transparent tissues one could see their insides, their gills vibrating, their organs sucking in the water, extracting microscopic quantities of oxygen, and expelling it again. One could see their body juices coursing through their muscles and could see what they had had for lunch. In a miniature way, these prawns were as gargoylish and as fantastic as any form of life on this earth. Each possessed eyes which glowed with crimson fire; these seemed to stare malevolently. Between the tiny fiery eyes protruded a long rostrum of razor-edged jagged saw teeth like some sort of undersea unicorn. And to make them even more unbelievable these crystalline prawns progressed, not forward as do normal creatures but backward jerking themselves along with flips of their fanlike tails.

These prawns, although fascinated by the light, were not wholly intent upon it. They were concentrating their attentions on the smaller crustaceans swarming about them. Darting here and there they seized the unfortunate animals that came within their reach, grasped them with their tiny clawed feet, transferred their victims to their jaws, and there shredded them apart. Their appetites seemed endless. One large fellow stuffed itself as fast as it could seize its prey. It was hanging in midwater only a few inches from my face. Through its limpid tissues I could see the parts of its victims filtering down into its body, where they resided in a blackish cluster.

Then, as though in retribution for their gluttony, the prawns themselves soon became the preyed-upon, for suddenly out of the blackness, sweeping only a foot or so above the bottom, came a group of fishes. Before I had time to move, these had sped through the cone of light, gulped down the feeding prawn, as well as a portion of lesser fry, and darted away again.

By this time the cold became too much to endure and with reluctance I let go of my sturdy but now thoroughly disheveled finger sponge, rose to a kneeling position, and was immediately swept out on the end of the line like a rag in the breeze. Then, shivering, I climbed hand over hand until the moonglow once more became visible. I felt the helmet raised from my face; I ducked out and in a moment sat quaking on the deck of the boat.

It was nearly midnight when I again slipped over the edge of the boat and prepared to don the cumbersome diving helmet with its eighty pounds of lead. In the distance the lights of a steamer bound for the ocean twinkled across the water and in another direction a lighthouse flashed fitfully. Closer by, a bell buoy rang irregularly in the doleful manner of all bell buoys; on shore a rooster, disturbed in its sleep or deceived by the moonlight, began to crow halfheartedly and then as abruptly ceased. The tide had apparently reached full ebb, for the spar buoys that marked the main ship channel were no longer pointing uniformly down the Bay but were lying at all angles. The wind, which until this time had been blowing softly, but steadily, was now quiet and the Bay had subsided into a great sheet of still silvery light. The last sound I heard before the waters closed over my head was the steady throb of the engines of the steamer as it passed down the channel and headed out to sea.

I slid down the line as rapidly as I could adjust my ears to the increasing pressure. Once beyond the aura of the moonlight, the phosphorescence was more brilliant than ever; the slighest movement caused the water to break into a blaze of sparks. In less than a minute I was on the bottom and this time I was able to stand upright, for the tide had ceased. In my right hand was a bag of loosely woven mesh and in the mesh to attract whatever fishes might be about were the crushed carcases of several blue crabs which had been brought along for the purpose.

The bottom was slightly different from that of the first descent, for we had lifted the anchor and allowed the boat to drift. The sea floor appeared to have been swept with a gigantic broom; it was clean and free of all debris. To my surprise I found it to be fairly alive with small flounders. They were invisible until they were touched, then in a flurry of sand they would break away and go fluttering off in the haze. I counted no less than seven in about thirty feet. Even in the light of the flash, which accentuated every little depression or elevation of the bottom, they were difficult to discern. They were of the exact hue and pattern of the bottom sand and so closely were they bedded down in the silt that they were indistinguishable. Once, in the dark, I actually put my hand on one. As I felt its quivering flesh surge up beneath my fingers, I was considerably startled.

The reason for the abundance of flounders soon became apparent, for this particular part of the Bay floor was alive with darting invertebrates and small fish. Apparently all these flounders had to do was to lie patiently in the sand and wait for the dinners that inevitably came within reach. With a quick snap they lunged upward, closed their sharp-toothed jaws on their prey and then snuggled close to the sand again to wait for the next morsel. The flounders furthered the feeling that other creatures were waiting patiently and quietly for their victims. In every place danger lurked for the small things of the sea. Even the sand was treacherous. So perfectly disguised were these flounders that later I saw a small blue crab settle on what it thought was a portion of the bottom. But it did not suffer under this delusion long, for it had hardly touched the disguised flesh when there was a flash of white and the tiny crab was crushed between rows of needle-sharp teeth.

The sudden and unexpected demise of the crab touched off a whole sequence of events which proved that there were indeed creatures waiting in the dark. For hardly had the crab been seized and swallowed and a few fragments gone floating down the tide when these lurking beings made their appearance.

The first came sliding out of the water from the direction of my right elbow. For the moment I did not know what it was. All that was visible was a pair of brilliant green eyes with soft lustrous irises. The eyes hung back in the haze and glowed iridescently. I was reminded of the famous Cheshire cat of *Alice in Wonderland* which appeared with a great smile. Then I saw that an inch or two beneath the eyes was a pair of large buck teeth, looking like the caricatured ivories of some Japanese militarist.

So fascinated was I by these mysterious eyes and teeth that I had forgotten the bag of bait in my hand. It suddenly became alive, and was quivering and jumping as though it would release itself from my grasp. A group of fishes were nuzzling and tearing at the mesh, as though frantic with hunger, and they seemed infuriated that they could not get through the strands and reach the luscious meat inside.

Among these fishes were several more pairs of the brilliant green eyes and I identified them as belonging to puffer or balloon fish, those curious creatures which, when scratched gently on their abdomens, swell to Falstaffian proportions. It is believed that this special ability to distend is a mode of protection. Their scales have been modified into rough prickles and when the fish is fully inflated it is as coarse as a piece of sandpaper. When menaced by predacious enemies they inflate themselves until they are too large to be swallowed with ease, and should any injudicious deep-sea dweller be so indiscreet as to tackle one it would find itself with a meal about as unpalatable as an unskinned pineapple.

I had hardly fixed my attention on the puffers and on the other fish swimming with them when without warning they turned and disappeared. I looked about for the cause of their hasty departure and at first saw nothing. Then off in the haze I made out a wide dark shadow which moved mysteriously just out of the range of my light, disappeared, and then came back again. From just above the cone of my light a great pair of wings swooped down on the bag of bait, nudged it, and then disappeared in the dark. As the wings went by I could feel a swirl of cold water. I allowed the bag to slip from my fingers and it fell and rolled on the bottom. In a second the shadow with its dark curving wings was back again and enveloped the bag as though to smother it.

There was something unpleasant about the way the animal hovered over the bait. Its flesh was dark and rubbery; from the base of the great wings protruded a long tail which stuck out stiffly behind the body and twitched as though in deep emotion. Less than midway down the length of the tail a sharp ivory barb stood out. The creature was one of the large-winged stingrays; and once when the tail whipped close I flinched, for I knew the ivory barb was capable of inflicting a painful festering wound. Through the dark-brown skin I could see the muscles straining. The beast lay there a long time, fluttering and quivering over its meal, and finally, when it had finished, it lifted the great fins and went silently away in the gloom.

After some hesitation about attracting another such companion, I rose to the boat, secured another bag of bait, and settled myself on the bottom again. To my relief no more stingrays put in appearance; instead I was visited by one of the world's most truly wonderful fish. I was seated with the bag suspended in front of me when softly out of the haze there slipped a long green undulating body. It came from above my head, twisted gracefully around under my arm, went behind me, and returned from the other side. Again it swept close, went curling off in the dark, then returned to the bait. It was one of the largest eels I have ever seen. Its tiny eyes were cold and cruel and its mouth was half open. It coursed around the bag, winding back and forth in figure eights. In contrast to the ugly spectacle of the winged stingray, there was an unsuspected beauty about the creature. Eels are not normally thought of as beautiful and when seen in the oily bilges of fishing boats or on a fisherman's hook they are indeed unlovely. But in the depths of the Chesapeake, in its own environment, this great eel was a thing of exceeding grace. Its very sinuosity was expressed in smooth sweeping lines, it moved without effort and assumed at all times only the most flowing of curves. But its real beauty was not in its motion but in the hitherto unsuspected iridescence of its soft silken skin. In its underwater home there was no evidence of the slime, or the dull-green hue so commonly associated with eels. Instead it was of a soft lustrous glow, leaf green above and pearl pink below. This pink altered in tone as the eel moved its coils. One moment it flashed pale-lavender fire, next delicate and evanescent yellows, then back to lustrous pinks. In its peculiar serpentine way this eel was as savage as the ray, but its savagery was ameliorated. Like the ray, the eel kept persistently at the bait, kept worrying it, tearing it apart bit by bit until it was gone; then as sinuously as it had come the fish wound gracefully away and I saw it no more.

When this second lot of bait was gone I was content merely to sit awhile and watch. For a time there was no movement, then there began filtering past the light small groups of transparent slim creatures which did not pause; they entered the zone of brightness, glistened for a brief second, and then wriggled out of sight again. So clear were they that one had to peer carefully to see them at all. They were only two or three inches in length and except for their glassiness and size were miniature counterparts of the eel that had just left.

Then I realized that these small wormlike crystalline beings were elvers, immature eels finding their way back to the rivers, to the streams and ponds from which their parents had come nearly two years before. For a long time they passed, hurrying on their way, about to finish their amazing journey from the depths of the ocean midway between Bermuda and the West Indies where they were born in the darkness and emptiness of the broad blue Atlantic. For many months they had been struggling through the watery wastes of the interminable Atlantic, swimming ever nearer the coasts, to fresh water. They were near journey's end and so impelling was their desire that the light did not attract them or turn them aside.

Minute after minute they poured by, at first a few dozen, finally in swarms of many hundreds.

I watched them until the cold became too much and chilled flesh begged for relief. Hand over hand I pulled myself to the waiting boat and to the open air beneath the stars. From the deck there was no sign of the life beneath, the Bay lay still and calm and dead. Only the buoys in the ship channels gave hint of the surge of the sea. They were all leaning, pointed one way. Like the eels below, they were oriented toward the dark land to the west. The tide had turned and was flowing in.

CHAPTER 2

LIFE BEGINS IN THE CHESAPEAKE

In the beginning God created the heaven and the earth. And the earth was without form, and void; and darkness was upon the face of the deep.

OF ALL THE CHAPTERS OF THE Sacred Script the most engaging are those of Genesis. In few other works of literature, religious or secular, are there bound together so many provocative and entrancing phrases. The most compelling is the very first—"In the beginning."

In the beginning, so the beautiful lilting words tell us, the earth was without form, and void. Physicists and astronomers speak learnedly and profoundly of the beginning of things, of electrons and energy, and of whirling nebulae. I have never seen an electron, and nebulae are vague swirls of light in the evening sky. Electrons, in truth, remind me of certain stuffy university classrooms rather than of the events that must have occurred on the last day of chaos, just before the early hours of earthly creation.

It is not the purpose nor the intent of these paragraphs to stimulate a religious-scientific discussion. Instead, and more properly, the function of these statements is to bring to attention an afternoon spent on the Chesapeake Bay near Cove Point, Maryland, where I once saw a wholly understandable and tangible demonstration in miniature of the marvel of genesis.

Scientifically, the demonstration bore no true comparison to the actual or even the probable, but it is not always well to reduce thought to cold scientific fact; for metaphor and parable are as much an art of intellect as of literature. The mere persistence of the allegory of genesis through several thousands of years affirms this statement.

On this particular afternoon a dense fog had descended over the waters and had enveloped the bordering land. When the fog had suddenly fallen, I was a quarter of a mile out in the Bay in some shallows where I had waded while observing the feeding of long arrowlike needlefish. Absorbed in their antics, I did not look up until the beach was veiled and I found myself in the center of a narrowing circle of mist. Presently I became conscious that I had forgotten in which direction the shore lay.

I stood still listening for some hint of land, chirp of bird or rasping of a grasshopper. There was none. I was enveloped in a humid, pearly, luminous yet opaque blanket to which there was neither shape nor border, substance nor tangible limit. Such indeed might well have been the prelude to creation. In the beginning there was—soft, gray, empty space.

For a time it was pleasing to be isolated, detached from the world, and I stood still, waiting. Then presently the air began to stir—faintly at first. The fog grew thicker and heavier and the wind faded again, whispering. In a similar manner might have begun the first stirrings of existence, a slight breathing, a momentary rustle among the waste.

All over my body, on the sleeves of my coat, on the hairs of my head, an interesting thing began to happen. Out of the air tiny droplets of water were forming, growing momentarily larger until some were a full thirty-second of an inch in diameter. Each droplet was a perfect sphere, unmarred by any blemish or distortion; each was poised lightly on the objects on which they were resting. Then I noticed that the air was full of these moisture drops; they were the fog itself.

These spheres were perfection. Unbelievably transparent they reflected the wan light with a lustrous brilliance. Clarity and purity was their characteristic. It is said that the Buddhists consider the morning dew their symbol of the ultimate in chastity. This gives their prayer formula "Blessed is the jewel in the lotus" a meaning understandable even to Western minds.

The sphere is the most perfect geometric form. It is the shape of the earth, of the sun, and of the stars. Curiously, it was taken as the symbol of the divinity by the first being of any consequence to conceive of monotheism, a certain strange character about whom we know little, the Pharaoh Akhnaton of Egypt. Also, the sphere is the shape of the primitive ovum, which is to say it is the form of life itself.

Out of the nothingness of the fog had proceeded in a few moments the contour of the earth and the heavenly planets, a symbol of godliness and of the prime form of life. And the allegory of genesis was complete when a moment later I dipped my hand in the salt water and very gently lifted it out again. Enclosed in my dripping palm was an exact counterpart of the crystal-clear water spheres. The object was many times larger than the fog drops and it quivered slightly but to all appearances it was exactly the same, pure, clear, and gleaming.

That I was able to do this was nothing extraordinary but an event that can be duplicated at any hour in the Chesapeake in certain seasons of the year. For the glistening bit of life in my palm was the quarter-inch body of one of the *Bougainvillia* jellies, transparent tiny animal wraiths which people the Bay waters, so clear in texture and so filmy in structure that one can swim through thousands of them without being aware of their existence. They are the free-swimming form of an abundant Chesapeake hydroid. They occur in the Bay at times literally in the millions.

These *Bougainvillias,* like most of the jellies, are ninety-six per cent water, four per cent clear tissue. But there is a subtle difference between them and the similar-appearing fog drops—the ability to reproduce their kind. If I could define this difference, I could, like Tennyson with his "flower in the crannied wall," tell what "God and man is."

No one knows how, why, when, where, or under what

conditions life first began. We can trace the evolution of animate existence back to a certain point. The record of geology is clear through the ages of the Paleozoic; the ponderous march of the years is recorded in the layered rocks of the Devonian, the Silurian, and the Cambrian. The mailed and armored fishes and the primitive sharks were preceded by the shelled mollusks, and by the even more fantastic trilobites and all the creeping crawling host of invertebrates. But beyond these the trail becomes vague and ill defined. In the old crumpled, twisted, contorted, and compressed archaic rocks, in the hard schists, granite, and marbles of the Huronian and Laurentian periods, the evidences of life are few. Some layers of black graphite, veins of brown iron ore allegedly deposited by organic action, and half-fused bits of white limestone indicate that perhaps the protozoa swarmed the ancient seas.

But somewhere in the archaic ocean or in the moisture of some warm, steaming barren river, or in a still, lifeless pond, or even perhaps in the crevice of some water-drenched boulder, the incredible happened. A droplet or cell-like globule of protoplasmic chemical substance quivered slightly, began to pulsate gently, and slowly split in two—and the long slow progress toward man and the mammals was begun.

I like to believe, and I suspect, that this dawn animal, this precursor of all living things, was like the glistening fog drops, or that it appeared as the spherical body of my captured *Bougainvillia* jelly, clear, lustrous, and sparkling.

Gently I lowered the frail body of the tiny jelly to the surface, saw it spread its limpid tissues and float off into the green Bay. So indeed the first creature might have done, yielded its newly created substance to the waters, to go and gather whatever nourishment the unpeopled seas of that ancient time had to offer and to impart the gift of life and reproduction to its multitudinous successors.

The subject of beginnings is an intriguing one. An awkward boy named Napoleon Bonaparte became an emperor instead of a prosaic writer of stories because the criticism directed to his first serious essay—with which he hoped to win a prize—so stung him and so defeated his youthful literary aspirations that in humiliation and anger he turned his thoughts to other endeavors. But for the curt words in a letter of rejection the tragedies of Austerlitz, the Nile, Marengo, Borodino, and finally Waterloo might never have occurred. The bloody trail that started in Corsica and led to the mirrored and gilted halls of Versailles and terminated on the lonely island of St. Helena may have had its beginning in a few scribbled words on a scrap of paper.

In a similar sense, Joseph the shepherd became the favored of Pharaoh and was made the keeper of the treasure of Egypt because his father gave him a many-colored coat. The Crusades spilled out of Europe in seven great waves of violence because, paradoxically, a very gentle man once walked the roads of Galilee. Newton is said to have formulated his most famous physical doctrine when an apple fell out of a tree.

The progression from apple to the law of gravitation is not so fantastic. Certainly these transitions are no more strange than those exhibited on every hand by the demonstrations of nature. Butterflies begin as creeping caterpillars; the golden blossoms of dandelions are first wisps of floating white silky down; roaring waterfalls have their genesis in clouds; and dragonflies come from stagnant, algae-covered pools. Only familiarity causes the vast gap between an acorn and an oak to be overlooked.

Consider, for example, the beginnings of those strange sea creatures, the sponges. Superficially, there seems nothing more prosaic or dull than a sponge; its chief characteristic is to be full of holes and to have an ability to absorb prodigious quantities of water. Yet for diversity and strangeness in their modes of creation the sponges are surpassed by few other animals.

Indeed, sponges are among the earth's most unusual creatures and they have long been a source of puzzlement. Their exact place in the scheme of things has never been fixed and there is no agreement among biologists as to precisely what they are. Unlike other, more familiar animals, they have failed to develop a head or a brain or anything approaching one; they have no mouth or digestive tract, no stomach or lungs or, for that matter, no other ordinary organs.

Yet, for all their failure to achieve a definable status, and although they seem fated to spend their lives rooted to rocks, piles and shells, stiff and immobile, they have succeeded in maintaining themselves for some millions of years. No small amount of credit for this long survival is to be attributed to their intriguing and varied reproduction.

The sponges of the Bay are not conspicuous or large like bath sponges, but they are legion. Nearly every wharf piling, every clump of oyster shell, every rock, even the bottoms of boats and ships are spotted with their soft yielding forms. Some appear as mere blobs of brown or reddish substance, others assume the shape of encrusting moldlike slime, a few actually drill holes in shell, rock, or wood and live there like hermits in caves; still other types grow upward in the sand like plants with branches and stems. The more adventurous fasten themselves to the backs of turtles, the carapaces of crabs and even certain quite old and moss-covered fish and thus have free transportation. For the most part, however, they lead quiet and sedentary lives, doing little more than allowing gallons of water to pass through their sievelike bodies and preparing themselves for the task of making new sponges.

It has been held that the most important function of living, indeed its very purpose, is to create more life. In this the sponges are paramount. Most creatures reproduce by division, or by producing "buds" which separate from the parent and go forth as new animals, or by sexual means. The sponges, not content, like ordinary beings, with any one of these systems, utilize all three, and have a fourth besides. Then, as if this were not enough to ensure their survival, they also possess remarkable powers of regeneration.

So tenacious of life are these creatures that it is almost impossible to destroy them except by removal from their element, by drying or by freezing. Chop them in little pieces, drop the fragments into water and, in due course, unless they fall in unsuitable places, the pieces will grow into new sponges. Crush the rubbery tissues through the finest silk mesh, pulverize them; it is probable that much of the strained or mangled flesh will survive.

Resurrection after death is a theological idea which has persisted since the time of primitive man. Nearly all the old religions have woven doctrine and dogma about the principle and have offered legend or scripture to support it. The ascension of Jesus, the monthly lunar rebirth of Ashtoreth or Selene, the carrying of the heroes to Valhalla, and the entrance of the Indian to the happy hunting ground are illustrations. Except possibly for the earthy reincarnation of the Hindus, all these ideas are on an abstract or entirely metaphysical plane. It remained for certain of the sponges to put into regular practice the creation of literal life out of death.

For this is the habit of some of the brackish and fresh-water sponges and it is their final provision for the onslaught of unfavorable conditions with which even their powers of regeneration cannot cope.

When the time arrives that the struggle for life becomes too much to bear, just as a dying tree will quickly put forth blossoms to ensure the formation of seed, so these animals make provision for the future of their kind. Within the failing tissues are formed strange bodies called gemmules which are equipped with all the needs of life, with cells to nourish and to create. The bodies are encased in a tough membrane for protection against the time of need.

Then, when the parent dies, when the old flesh is falling apart and festering, the gemmules lie quiet in the disintegrating tissue until the process of dissolution sets them free. Out of the dead animal they roll to settle on the bottom of the Bay and then at the propitious time the gemmule opens and the newly born cells stream out to rearrange themselves in the cool green water and begin the complicated business of making a new sponge.

Other forms, those which frequent more salty waters, vary the system a bit further. For these there is no patient waiting within a rotting corpse. The gemmules of these do not need a hard, tough shell for protection; instead the surface acquires hundreds of little hairs which by waving rapidly in unison move the larvae through the drifting waters. After swimming about for a time, the hairy midget attaches itself to a suitable object, a shell, a rock, a wharf piling, the bottom of a boat, and in time becomes a sponge.

A number of the sponges are truly sexual—but in the perverse way of sponges they are unlike other animals in this respect too. They may be all male or all female; more likely they will be both in one individual. Strangely, although they may produce both sperm and ova in abundance, there are no special structures to provide these organisms.

The sperm cells come into being by diverse means. In some cases the entire interior of a sponge may become transformed into a mass of spermatozoa which break forth and stream in large numbers into the Bay, where they drift freely until sucked into the body of another sponge. In other instances the sperm are carried and cared for by special cells which protect and nurse them until an egg is reached.

But whatever system is utilized and however the union of sperm and ova achieved, the egg develops, forms a tiny sac covered with hundreds of little lashing hairs, squirms its way through the body of the parent, and emerges into the water. The beating hairs, like oars on an old Roman galley, scull the creature on its way. In a few hours, if not devoured by some hungry being or hopelessly lost in a waste of water, it fastens itself to some solid object, glues itself securely in place, and busily goes about the important task of making another sponge.

The sponges are unorthodox and fail to observe the ordinary regulations of being born. Unorthodoxy, however, is the mode of many of the Bay's animals and the sponges are no more peculiar than the pipefishes. These latter beings reverse all the accepted procedures and practice, in true Thorne Smith fashion, a system of turnabout. For it is the male that carries the eggs, gives birth to the young, and nurtures the babies until they are able and ready to care for themselves.

Pipefishes resemble sea horses which might have been pulled through the proverbial keyhole. Indeed, they belong to the same family—the *Syngnathidae*—which is a ponderous way of saying that they are fishes characterized by being covered more or less completely by bony plates and by possessing a snout equipped with the smallest of mouths. Like the sea horses, they feed on tiny crustaceans, microscopic larvae, and other minute beings which they suck in by a clever vacuum-cleaner arrangement. Unlike the sea horses, however, they spend the greater part of their lives in a proper fishlike horizontal position and they still retain the dignity of a swimming tail; although it is to be admitted that the proportions of the tail are a little ridiculous. The pipefishes are further distinguished by having no true gill flaps; instead, the gill opening has been modified to a small pore. At best they are poor travesties of fish. They swim badly, are stiff in the joints, and their movements are silly and futile. But in spite of all these handicaps they are highly successful beings. It is my conservative estimate that the pipefish population of the Chesapeake Bay is in the order of 125 million individuals. Few persons appear to be aware of their existence, yet every patch of seaweed contains a legion of them and they occur in all parts of the Bay from fresh water to the Virginia Capes.

How the incredible change of relationship between the sexes ever came to be no one knows. Perhaps some exasperated female pipefish a thousand generations removed turned up her figurative nose, informed her incredulous spouse that she was no longer worrying with the housework and the children and angrily dumped a bunch of eggs in whatever served her husband for a lap. Actually the turnabout was the result of a slow evolution, for both male and female have been structually changed to make the alteration possible.

In the cool green waters of the Chesapeake any time between April and October, the spawnings take place. In the shallows between the swaying long tendrils of seaweed the males and females meet. With awkward gyrations and vibrating fins their bodies move close. At the appropriate moment, when the *Syngnathic* emotions have reached their peak, the female extrudes a long tube which she places in a pouch on the male's abdomen. Along this tube flow the as yet unfertilized eggs. One by one they are laid in place until several hundred fill the cavity. Here they are fertilized by the swarming, searching sperm cells.

This is a thoughtful provision to ensure that neither sperm nor eggs will be lost or dissipated by the swirling tides. Safe within the male natal cavity, the marsupium, the eggs begin their crowded development. Her purpose accomplished, the female swims off to follow her own devices, a free woman, so to speak. The eggs incubate and hatch inside the pouch. In time the lips of the swollen marsupium open and the fully formed young pour out into the misty green Bay that is to be their home for the remainder of their lives.

The sea creatures are ushered into life in the soft pastel liquid of the Bay, where the temperature changes but little from hour to hour, where the light is filtered and diffused, and where day differs from night only by the gradual lightening or darkening of the water. Save for the slow seasonal changes, the world is forever static; stillness and silence are the rule, only the motion of the waves and the change of the flowing tide denote the passage of time.

How different the scene of the nativity of a baby tern.

I recall the day some years ago when I first attended the hatching of one of these delicate and graceful sea birds. The scene was Cove Point in Maryland not far from the mouth of the Patuxent River. It was the 28th of June and I was sitting on the sand with my back propped against a whitened barnacle-encrusted log which had been washed up on the beach by some long-forgotten storm. In that year Cove Point was not yet touched by the events that later were to ruin it forever; it was one of the loveliest spots on the Bay. To the south the tall blue cliffs of little Cove Point rose a hundred feet in the air and they were broken in places by dark ravines which terminated abruptly on the Bay front and spilled masses of green vegetation over the cliff faces.

A semicircular white beach swept from the base of the cliffs and enclosed an arc of blue water fringed with a fine line of gleaming breakers. Back of the beach a strip of golden beach grass waved gently in the wind; beyond this a grove of cedar trees gave an almost parklike effect as though someone had planned an exquisite Christmas garden and planted them on purpose. Close by and partly hemmed in by these cedars was the Cove Point lighthouse, a slender column of white stone situated almost on the water's edge. Beyond the light a long spit projected into the Bay. The spit was made of yellow sand, wave-strewn pebbles, and seashells. At the tip the tides swirled past in visible eddies and waves chewed ceaselessly at the shifting tenuous sands.

On this nethermost point there was almost no vegetation. Only a few scattered blades of beach grass survived the continuous glare, the salt water, and the heat. Even as I looked, the horizon was distorted by heat waves rising in quivering curtains. The reflection from the water was blinding, for it was very calm and the sun was near the zenith. The sand burned to the touch and when I idly picked up a rounded pebble it was almost too hot to hold.

With my glasses I scanned the point carefully. Through the powerful lens I could see each individual oyster shell, each worn pebble, the little frayed pieces of sea grass, the deserted and whitened carapaces of long-dead crabs, and a host of fragments of broken bits of wave-washed wood, the bones of dead fish, drifted sea wreck, and the flat, still bodies of stranded jellyfish. At first I could not find what I was looking for but presently on the hottest, driest part of the beach I found what I sought.

Sitting on the bare sand was a diminutive white shape. It was the body of an adult Least Tern. Except for the black cap on its head and the dark line running through its eye I would have passed it by. Through the glasses I could see it was watching my movements. Carefully noting the exact spot, I got up and moved closer.

Almost instantly the tern was off, and was winging excitedly back and forth uttering the most plaintive cries. With the first notes three or four other terns which I had not seen rose from the beach and added their voices to the clamor.

With a flash of wings the bird dove at my head, screeching as it plummeted. The air from its wings brushed my face and slightly ruffled my hair. Again and again the bird attacked, sheering off just in time to avoid actual contact. But I paid no attention and kept my eyes on the beach where I had first seen the creature. When I arrived at the spot I found, as I expected, two small cream-colored eggs speckled with brown. Beside the eggs and partly surrounded by broken pieces of shell was a newly hatched chick. It was still damp and grains of sand clung to its sticky body. I took one glance and quickly retreated to my log.

The time was 11:45 A.M. Once I was out of the way, the parent speedily returned, and through the binoculars I could see it shuffle over its offspring, after first picking up the broken shells and dropping them several feet away.

At precisely 12:27, forty-two minutes later, I again visited the depression in the sand. The two eggs were still intact but a great change had taken place in the chick. It was no longer slimy and wet nor did it appear to be bedraggled and gaunt. Instead, it was a clean, dry, beautiful ball of fluff. Out of the yellowish, sand-colored, mottled down appeared a pair of bright dark eyes. It was hunched well down into the hot sand and in response to the frantic warnings of the screeching mother was immobile, "frozen" in the most concealing position. When I picked it up and moved it a foot or so away, it scampered back into the nest and again froze into position.

Think of the transition taking place during that forty-two minutes! For fifteen days the little embryo had been curled up in its shell of calcium, unseeing, insensible, saturated with amniotic fluid, immovable, imprisoned, impassive except for a few prenatal twitchings and squirmings.

Then, with the breaking of the shell, its cosmos was

altered from an oval still cell where nothing changed or moved, to a blinding world of brilliant light, where one's damp, sticky body was suddenly deposited on hot, gritty sand, where only a few yards away a seething liquid green substance beat against the earth and emitted a thunderous sound with every surge; where a vast blue space existed above broken only by strange, white shapes which slowly drifted away.

It is not good natural history to place human interpretations on the reactions or thoughts of animals and birds. But I suspect that the sudden advent of a baby Least Tern into its shining realm is accepted unquestionably by the chick as good and natural. Implanted somewhere and somehow in the growing embryo, or, if you prefer, in the genes and chromosomes of the sperm and ova of the parents, is a fund of instinct and knowledge of what is proper for baby terns to do.

It is certain that in a few minutes, once the bright, beady little eyes are open and the full scope of its hot, gleaming world has been absorbed, the chicks are doing those things which fall to the lot of infant terns as though by long experience. The short, tiny legs, which had been cramped within the confines of a thirty-millimeter shell, perform their functions perfectly, almost without trial. Think, in comparison, of the tribulations of a human infant first learning to walk, of the interminable falling and getting up again, the weak, ungainly, drunken stagger that marks the human progress to ambulatory success. Within an hour or two of birth Least Tern chicks can scamper all over the beach, will respond to calls of their parents, and will lie motionless until danger is past or until a change in tenor of the call signals permission to relax.

The birth of a baby tern is an ushering into a world of danger and of hostile elements. Hungry animals prowl the beaches, birds of prey hover in the skies; the tides and the waves are never far distant. But it is a clean bright world and in exchange for the hazards of being born exposed and unsheltered is the possession of a guiding instinct, a coat of well-camouflaged down, and the attentions of a pair of fiercely devoted parents. Most of all there is the compensation that within a few days or weeks this clean, blue world of golden sand, gleaming surf, and sparkling waves will be free for roaming at will, with the privilege of diving in the green shallows after delectable minnows or, simply, as is the way of sea swallows, to go hurtling over the wave crests on the wings of the wind.

The terns are born in summer heat and in the midday glare and their new-found world is all light and activity, sunshine and blue sky. This is in contrast to the beginnings of other Bay creatures; for certain of these are born in the dark and blackness, away and sheltered from the world of the open air and the time of their birth is in the late winter when the Bay air is cold and raw. They come into being in the heart of the marshes and the swamps. In the places where they are born the air is damp and close and heavy with a thick animal smell. The odor of musk saturates everything with a blanket of scent. Beneath the musk are other odors: the pungent smell of wet mud, the earthy scent of crushed reeds, a suggestion of salt mixed with the aura of dead fish such as at times lingers above the bilges of old boats, the cleaner smell of grass, dried leaves, and dissolving vegetation.

The darkness is unrelieved and only movement indicates the presence of life. There is a splashing of water, of drops falling on an invisible surface, a slithering, scraping sound and the hiss of breath being gently withdrawn and expelled. In the distance, and strangely muffled, are other noises. Most powerful is that of the wind, although it is muted and soft. It is blowing heavily, for it is early March and the gusts produce wailings and whistles which rise and fall in a succession of crescendoes and diminuendoes. Peculiar patterings accompany the blasts, the scraping and rasping of long-dried vegetation, the sound of reeds bending and recoiling to the pressure, clattering and banging one against the other. In the distance and heard only occasionally is the sound of ripples, little lispings of water as these, pushed on by the wind, break against the mudbanks or lose themselves between the water grasses. Even farther away, and more suggested than heard, is the heavier pounding of surf—of Bay water breaking on a beach.

For the scene is the heart of a muskrat house in a Chesapeake swamp on the natal night of a new generation of muskrats. The house stands half submerged between the reeds. Old ice lies piled about the structure and partly covers the submerged tunnels and runs that radiate in all directions.

These muskrat houses dot the Chesapeake marshes by the thousands. They appear as mounds of reeds, roots, and mud from six to eight feet in diameter and about three feet high. As is characteristic of the houses of the Bay country, they are somewhat flattened on the top and are covered with a layer of dry cat-tails and marsh grasses.

They are—at least from the viewpoint of the muskrats—wonderful places in which to be born. Snug and safe from the wintry blasts, secure from predators—from the nocturnal, soft-winged owls, the sleek, lithe, blood-hungry weasels, and the foxes, guarded from the evils of the outer world, the baby muskrats come into being. Snow, rain, hail, ice, and lashing gales exist only as faint sounds or distant effects beyond the sturdy, insulated walls. Only three enemies are to be reckoned with: the grizzled raccoons, which alone among the preying mammals have been known to tear open the houses; rarely, the lurking treacherous tide which once or twice in each year or two, pushed on by a lashing northeaster from the outer ocean, forces the water higher and higher, fills the winding courses of the swamps, covers the roads and submerges the snug, dry houses in a wet flood; and last, the destructive fingers of man.

The natal bed of the muskrats is a small shelf lined with soft grasses and the down from the neighboring cat-tails. The shelf is located a few inches above the normal high-tide line and not far from the water-filled tunnel entrances. Occasionally three or four of these chambers will be formed in a single house, commonly only one. Here in the darkness, twenty-nine or thirty days after conception, the heavily laden mother comes for her hour of labor. Alone, for the male has long since departed or has been driven off, she undergoes her travail, silent except for heavy breathing and ratlike squeakings of pain—we do not exactly know for no one has ever been present at one of these rodent accouchements. One by one the hideous, squirming, slimy, wrinkled travesties of muskrats are laid quivering and pulsating on the bed of grass. In the fetid darkness, they lie between the mother's legs, damp and sodden. Their eyes are blind and closed, but it makes no difference, for there is nothing to see. But in their infant way they sense the mother's body heat and squirm closer, wriggling on flabby legs, more like serpents than mammals, until the rubbery tissues of the mammary glands touch their ugly, toothless heads. Then, warmed by the contact and stirred by strong mammalian instinct, they open their diminutive mouths and search blindly for the teats with their life-giving milk.

Tired, weak from labor, exhausted by the pain of birth and the effort of cleaning her young, the mother curls about her babies and dozes off while outside in the cold night the wind stirs among the reeds, whispering about the broken stalks of the wild rice and causes the water to darken as the surface fractures into a million facets.

Thus, with a sponge, a bird, a fish, and a mammal, we complete this survey of genesis. With these four representatives an indication has been given of the complex and intricate systems by which the multitudinous creatures of the Bay perpetuate their kind and ensure the continuation of their species. The marvel of genesis is the most common of the world events and the least understood of its mysteries. Whether the occurrence is the birth of a babe, the unseen ejection of a naked mammal in the dark recesses of a half-submerged den, the cracking of an egg on the gleaming sand of some wave-swept beach, the release of a microscopic fish in the flood of green waters, or the outpouring of billions of wriggling spermatozoa from the flesh of a slimy brown sponge, it is the event that most closely approaches the answer to the most intriguing question of all.

In the beginning there was . . .

CHAPTER 3

LIFE PASSES ON

To write of the beginning of things is relatively easy. Birth is a phenomenon which is accepted as natural and good. The advent of a child is hailed with joy; the coming of puppies and kittens is a matter of pleasant interest; even the occurrence of squirming tadpoles in a previously empty pond is considered a reasonably happy event. A whole science and a considerable literature has been built up about the wonders of creation. Colleges have endowed classes of genetics and our inquiry into the business of being born has been studious and serious.

To frame a chapter on death, however, is something else. One is instantly suspected of being in poor taste and the very nature of the subject poses a literary hazard. Only theologists and the writers of detective stories appear to have entered this field successfully and with impunity, which is a pity because the topic is a fascinating one. Natural science, in particular, has neglected this aspect of organic existence. Our libraries bulge with works on the identification of plants and animals, with ponderous treatises on their behavior, their anatomy, with long monographs on evolution, adaptation, and similar themes. But there is no shelf on the death of animals, nor volumes to fill such a shelf if a librarian could be found sufficiently indiscreet to provide one.

At the risk of committing literary hara-kiri, I plan to devote an entire chapter of this book to death and I propose that the occurrence is a reasonable and interesting function, that no approach to an understanding of life is possible without a rational acquaintance with the phenomenon.

It is never quite possible to regard the death of human beings with equanimity. The emotional response is too deeply rooted to permit calm intellectual inquiry. Anyone who has lost a loved one or who has experienced the doleful half-pagan ritual of a modern funeral with its somber trappings and display will appreciate this assertion. It is only in the realm of nature that one may contemplate death with a comparatively unbiased viewpoint.

To attempt to define death is like trying to define life; there is no answer to either query. A creature is suddenly born; it spends so many minutes, hours, or days breathing, growing, catching and assimilating its food, indulging in the activity of its kind, providing for the duplication of its species. Late or soon it is devoured, is killed, or simply ceases to function. Why it is created and why it is cast aside has never been satisfactorily explained, even by the theologians. Some few animals, those of very simple structure, like certain of the protozoans which are created by fission, achieve an immortality of a kind as they divide again and again and thus pass the gift of life along, ever creating new individuals. But for the mass of animate existence there is a positive termination, an exact point beyond which is only decay and dissolution.

Every Chesapeake beach is a catalogue of organic failure. The area between high and low tide is strewn with the wreckage of a multitude of lives. The shells of mollusks lie broken and abraded in the sand; the fragments of fish bones protrude between the rolling grains; occasionally flashes of shining silver denote the presence of scales detached from some finny body which once moved in the cool green depths; logs spotted with barnacles bleach in the hot sun; and the lacy patterns of colonial bryozoans filigree graying strands of withering seaweed. Between the windrows of piled-up algae are countless remnants of crabs, whole carapaces intact with spines and still colored with the hues of life or perhaps turning red with the action of the sun, single legs and claws, pieces of joints, flattened flippers, even individual many-faceted eyes, sightless, dull, and unseeing. Mixed with these are the broken portions of other crustaceans, the transparent armor of shrimp and prawn, the empty husks of copepods and isopods.

Nor is this collection of destruction limited to the things of the sea, for the beaches are strewn with the remnants of the life of the air and the land. Here in one great heterogeneous assemblage is stored the evidence of the universal tragedy. Sodden wings, loose feathers, the broken, crushed elytra of beetles, the fragile tissue of butterfly wings, parts of grasshoppers and other bugs, the seedlings of plants that somehow fell into the salt water instead of the fertile soil for which they were intended, wisps of the down of the milkweed and dandelion drifted too far on the wind, clumps of grass and flowers torn by the waves from their abiding places and floated across the Bay, whole trees and saplings washed out by the floods of the rivers, and other wreckage everywhere litter the sands in profusion.

But, while the Chesapeake beaches are the depository of all those things which have passed their time of usefulness or which have been struck down or have been overwhelmed by circumstance, they are in no sense charnel places. No smell of death persists, save perhaps a slight aroma of fish and rotting seaweed, and this is not unpleasant. The hot sun and the ministrations of the scavengers keep the sands clean and sweet. No sooner has a dead thing come to rest on the grains than the vultures are there to clean it away. These and the crows dispose of all the large objects; the beach crabs, the carrion beetles, the maggots care for the rest. Those remains not suitable for the ghoulish tastes of the scavengers are rapidly dried by the sun into hard-baked nonodorous fragments of tissue; these by the resolving power of their chemistry soon return to the elements.

I can recall only two exceptions to this general rule. Once in 1936 while sailing along the shores of St. Mary's County above the mouth of the Potomac River I came upon a stretch of beach about two miles long which was fetid with the odor of decay. Some great underwater change had occurred far out in the Bay, for the sand was piled knee high with long dank masses of seaweed. The weed filled all the space from the water's edge to the pine forest in the rear. Tangled in the mass were the corpses of innumerable fish, crabs by the dozens, and great wet blobs of transparent substance from hundreds of enmeshed jellyfish. The tragedy was caused by a disease of the eel grass which started south of Florida about 1932 and in a few years traversed the Atlantic via the Gulf Stream. For over five thousand miles along the American coast similar beds of weed broke loose, carrying to premature death innumerable fishes and invertebrates which depended on the grass for shelter or sustenance. The poisonous disease also slew hundreds of water birds, particularly certain ducks and geese which fed upon the weed. During the first years of the blight nearly ninety per cent of the migratory brant geese died before they could learn to feed upon substitutes.

Nowhere in the wholesale destruction was there any sign of life. Even out in the shoals the usual legions of minnows were absent, acres of putrefying vegetation and protoplasm had seemingly poisoned the water for yards about. Strangely, not a vulture or other bird was to be seen. It was as though the mass of destruction was too great for even these servants of death.

The other instance occurred in the fall of 1919 near the mouth of the Piankitank River in Virginia, when an enormous school of fishes, croakers, Norfolk spots, and other species, were trapped in shallows where they had gone for some unexplained reason. A rapidly falling tide had left them stranded, a most unusual occurrence, and their bodies carpeted the sand bars in masses of gleaming silver. Then when they were all dead the tide had come in again and washed them all ashore, where they lay piled in long gleaming rows. This time the vultures came by hundreds and feasted until they could scarcely fly. But there were so many bodies that even the carrion eaters could not cope with them and for days the beach stank and the hum of green bottle flies rose above the sound of the waves.

There is little waste and little loss, for the failure of one animal is success for another. Nearly all life directly or indirectly is predicated on the destruction of other beings; only certain primitive plants appear to be able to secure all their needs from earth, air, sunshine, and water. A considerable category of creatures are specialists in death. The most obvious and among the most successful of these are, of course, the vultures.

Two species occur in the Chesapeake region, the so-called Turkey Buzzard and the lesser known Black Vulture. This latter bird is not found, commonly, in all parts of the Bay country but lives in the southern portions from about the vicinity of the Patuxent River to the Virginia Capes. It is a much neater bird than the more familiar Turkey Buzzard and is distinguished from the former by its black head, shorter and more rounded wings, and smaller size. But in their search for dead things the two are alike. On expanded, motionless wings they patrol the long expanse of beaches, the tidal flats, the meadows, and the marshes. Singly or in little groups of two or three, circling, soaring, gliding, they are an ever-present part of the Chesapeake background.

Once, on a sixty-mile cruise along the western shore of the Bay, I kept count of the vultures seen on the way. Never for more than a moment or two during the twelve-

or fourteen-hour trip was the sky free of their wheeling forms; they maintained a ceaseless vigil, endlessly scanning the earth for carrion. Each bird kept close watch on its neighbor, and when the vigilance of an individual was rewarded with the sight of some newly cast-up carcase and it began the long spiral to earth to claim its prey, so every vulture for miles around abandoned its portion of sky to glide in to share the spoils. Thus by a communism of watchfulness is food assured. Out of death, by virtue of their ghoulish habits and passive natures, the vultures glean an unenvied but secure place in the scheme of things.

The names of death are legion and its forms are infinite. It may be swift and sure, lingering and painful; some of its aliases are accident, hunger, carelessness, cold, overpowering heat, ill fortune, disaster, and old age. Death is the form of an osprey hurtling out of the sky on the half-submerged body of an unsuspecting fish; it is the silently flowing ebb tide that leaves the shape of a glistening ctenophore or medusa on an empty sand bar to wither in the sun; it is a sponge that fastens to an oyster and with the passage of time smothers its host in a rubbery mass of tissue; it is the tentacled body of a squid about to enmesh its prey; it is the wind that carries the body of a frail butterfly out over the Bay to perish miserably in a waste of water; it is a flood far up in the mountains of Pennsylvania or West Virginia flowing down the Susquehanna or Potomac that changes the salinity of the Bay water and causes hosts of creatures to die in agony; it is the approach of winter and the lowering of the temperature; it is the gradual silting of a swamp or the washing away of the beach during a storm; it is the accumulation of time and it is the flash of gleaming teeth and it is the stroke of ivory claws.

The actual death of animals is rarely witnessed, and the fate that overwhelmed them can only be inferred. I have often found the empty shells of tortoises in the pine woods with every bone intact resting in natural positions with no mark of violence upon them. They seem simply to have gone to sleep as though lulled by the whispering of the wind in the needles above. Some were young, some were obviously old; life departed easily and quietly, slipping away gently, imperceptibly.

The sea turtles are not so fortunate. During the summer they swim north on the Gulf Stream and wander into the Bay attracted by the deceptive warmth and abundant food. In the tropical land from which they come the temperature does not materially change and they are not prepared for the strange chill that creeps into the warm Bay water and slowly paralyzes their flesh. As the days pass from summer into fall, the water becomes more and more cold and the turtles' actions become more and more sluggish. Presently they can move no longer and cannot lift their necks to breathe. For all their bony protective armor they are helpless against this invisible piercing enemy. Their bodies drift ashore in dozens and the beaches of the lower Bay are often strewn with their carcases.

Death in the Bay usually does not come peacefully. Most of the creatures of the Chesapeake meet their ends violently, some few horribly. In this last category I recall a descent I once made beside an old abandoned wharf piling. The ancient wood was festooned with long streamers of lacy red algae which waved gently with the tide. Between the algae, the white houses of the barnacles clustered in close masses. The tunicates crusted with colonial bryozoans hung on the riddled pile in bunches like misshapen grapes. Between these in every available space were numbers of flowery, fawn-colored anemones. Their petallike tentacles were fully expanded and the light coming down from the surface shone through the amber translucent flesh. Against the somber green water these delicately hued animal-flowers acquired an aesthetic effect, an effect paradoxically heightened because it was subdued. The waving tentacles assumed only the most graceful positions. Their very delicacy and translucence belied their actual, sinister character. For some minutes I watched them and admired their subtle beauty. While I was inspecting this peculiar underwater garden a small fish swam by and circled idly about the piling. Slowly it glided between the threads of red algae, paused to nibble at some microscopic tidbit, and then hung motionless in midwater. With a quick motion I raised my hand to seize it. Startled, it turned and blundered head first into a set of the innocent-looking flowery petals. Instantly the tentacles coiled about the body, piercing it with hundreds of stinging, poisonous darts. The fish gave an agonized twitch or two as new tentacles wrapped about the tender flesh releasing more darts, and then it lay motionless, stunned into a contorted paralysis. Slowly the stilled body was passed from tentacle to tentacle toward the center of the anemone; the mouth distended and began to envelop the victim. All the while the unoccupied tentacles twitched and squirmed and slithered over the body of the fish. For sheer horror I have seldom seen anything to equal the sight; and I quite forgot that the anemone and the fish were only a half inch or so in length, that this was only a tragedy in miniature.

Panic and fear are the constant companions of death. Their favorite victims are the menhaden or alewives, those peculiar, vertically compressed fishes which haunt the Bay waters in legions. These creatures appear to spend their entire lives in a continual state of alarm, in a never-ending alternation between nervous calm and stark terror. They live in great schools, vast hordes comprising millions of individuals sometimes covering acres of the Bay's surface. They are plankton feeders and they subsist on the microscopic life of the Bay, which they secure by straining the water with its contained organisms through their mouths. By means of cleverly designed oral sieves they extract this minute manna and swallow it unchewed, as though it were a sort of living soup. Because of the semiliquid character of their food these fishes do not require teeth, so they have none. Thus they are quite defenseless; because of their helpless natures they have become the preyed-upon and are themselves the source of food for the underwater carnivores.

Their only means of protection is to gang together in enormous schools and by the very largeness of their numbers to create confusion to the enemy. It is probably a much more bewildering task to attack a mass of darting, scurrying forms fleeing in all directions than to track down a single individual.

To see one of these hordes of menhaden underwater is a never-to-be-forgotten sight. I have been completely surrounded by their massed golden-hued shapes. Packed in serried ranks, head to head, tail to tail, they have poured past my helmeted body in a deluge of finny forms. Nearly always they are in a state of high nervous tension. The slightest motion will cause them to break into terror-stricken flight; instantly the mechanisms of fear take hold of the entire school and the hysteria is communicated from one individual to another until the entire mass is transformed into a blurred deluge of frenzied fish. Then, inexplicably, the terror passes and almost instantly they will cease their dashings and once again move slowly through the water. Frequently minor attacks of nervousness will occur and send waves of temporary panic through local areas of the school.

Their nervousness is not unjustified, for always, just beyond sight in the dark-green water, is the menace of gaping jaws and needle-sharp teeth. Day or night, at any moment, powerful flashing shapes may burst out of the haze to start the attack. The enemy may come singly or in hordes; when the latter occurs the carnage is enormous. Blood and bits of mangled flesh fill the water, slicks of oil from crushed tissues float to the surface and lie in glistening yellowish streaks. There are times when acres of the Bay's surface is aboil with the surge of fins and the lashing of powerful tails and with darting crazed bodies leaping into the air to escape, only to fall helplessly in again. For miles around the sea gulls come hurrying to the scene attracted by the froth and the hope of food. In wheeling flocks they join the attack, screaming in excitement, plunging in to seize some tidbit and then circling to dive once more. Beneath the water the scent of blood and of fishy oil spreads in ever-widening circles, carried by the tide, attracting even more carnivorous schools of predacious rockfish, ravenous bluefish, small sharks, and a host of lesser beings.

The crazed, terror-stricken menhaden, with no defense, can only flee. Their safety lies in their vast numbers; the schooling habit that causes them to become ready victims is also their salvation. Theirs is a communism dictated by instinct; as in all communisms, the individual is sacrificed for the mass. Hundreds die in every attack but thousands escape only to become in their turn the sacrificed. Few menhaden ever succumb to old age; they are born to die violently.

Littered all over the beaches of the lower Chesapeake and lying in clusters on the bottom are hundreds of empty shells, the casings of oysters and other mollusks. At first they seem intact but on close examination many are seen to have holes bored through the shells as if someone had gone busily among them and drilled them with a fine bit. The holes are small, seldom exceeding one-eighth inch in diameter, and are smooth and polished. Each is evidence of the cause of the death of the mollusk that once owned the shell. Mollusks lead quiet sedentary lives. Inside their houses of hard lime they reside fortified against the terrors of undersea life, exempt from the danger of sharp teeth and rending claws. They have traded mobility and intelligence for safety and stability, but they have reckoned without their own kind. While they may retreat into their shells on the approach of an enemy and wait patiently for it to be gone, there is no escaping the carnivorous attentions of their cannibalistic brothers. For the tiny holes on the dead shells are bored by other mollusks, by little spiral-shaped creatures so small that the idea of one of these causing the demise of an oyster or clam fifty times its size seems unreasonable. Yet in their patient slow way these molluscan Davids overcome their shelled and armored Goliaths in a manner so horrible as to make death by swift attack seem desirable.

When one of these spiral gastropods finds another mollusk it attaches itself securely to the shell and begins its devilish work. Coiled inside the diminutive body is a radula, a sharp filelike device; this is extruded and touches the hard lime beneath. Back and forth it rasps across the shell, abrading it slowly away, a grain at a time. Also, there is evidence that the boring is aided by an acid secreted by a special gland.

There is no hurry, for the prey cannot escape; it is held fast in the house that was its protection. As the acid softens the lime, the file rasps patiently away, cutting little by little, ever deeper. If the teeth of the file become dulled new ones move forward to take their place. Back and forth, back and forth; layer after layer the calcium is removed. The minutes creep into hours, possibly into days. But, late or soon, the last fragment of lime is removed or etched away; a hole round and smooth lies above the tender quivering flesh. Down through this hole the hungry gastropod thrusts its proboscis, deep into the living animal. As long as the victim lives its tissues are torn out bit by bit, its body juices drained away. And when it has given up its final quota of substance, the loathsome little creature relinquishes its hold and slithers slowly away to find another victim.

This sort of thing has been going on for a long time if one may accept the evidences of paleontology. For miles along the western shore of the Chesapeake Bay above the Patuxent River and in other scattered localities great deposits of fossil shells are exposed to sight where they have been thrust by upheavals of ages past. These shells belong to mollusks of types now extinct and are believed to have lived in a softer, more tropical climate. In great strata they lie piled where the passage of the years has left them; or they are mingled with more modern shells in wave-swept windrows beneath the blue clay cliffs whence they have fallen. Almost every fifth or sixth shell has a neat round hole drilled through it and tucked here and there among the debris of these dead victims of thirty or forty millions of years ago are the shells of the tiny beasts that did the drilling. There is no evidence of what may have overwhelmed them in their turn. Disease, age, or other causes, whatever it was, they too lie buried with the accumulated silt and sand of a hundred thousand centuries piled over their remnants.

Among the shells, scattered through the layered strata, are the bones of giant whales, the ribs and vertebrae of porpoises, the carapaces of huge turtles, the teeth of crocodiles and sharks; some of the latter were so large that a man could have walked erect into their opened mouths. Even a few bones of birds have been found.

But of all the host of beings that must have lived only the mollusks have left their traces in abundance. Except for those few creatures which laid down their lives in favorable circumstances, which were quickly covered with the concealing, protective silt, all are gone. The sea that nurtured their bodies and received their dead carcases is gone too, so likewise are the tropical lagoons in which dwelt the crocodiles and the large turtles. Where the waters once lay still and quiet in the hot sun are now tall cliffs, rounded sandy hills covered with shady foliage and long green ferns.

There is one common destiny. Fish or amphibian, man or mammal, protozoan or crustacean, insect or serpent, all pay their ultimate tribute. The continual and repeated struggle, the battle to persist from moment to moment, from hour to hour, or from year to year serves only to stave off for a longer or shorter time the inevitable result. Yet, in a way, there is no positive termination. The individual perishes, disappears, is forgotten. But the stream of which the individual is a micro-portion continues. There is a very real link between grandfather, father, son, and grandson; their hands are joined in more than a figurative sense.

Thus, as the raindrop that falls in the Pennsylvania mountains loses its identity and becomes with its brothers a trickle down a fern-clad hillside, as it burgeons into a brook, a tributary, and becomes finally a portion of the broad Susquehanna River and, further, as it loses itself in the salt-tinged Chesapeake Bay, so the individual life merges into the current and makes the flowing possible. The concept of a life with its beginning and termination as part of a continuous and continuing process is more satisfying than the idea of seemingly purposeless birth and annihilation.

In nature there is no evidence of a life after death whether of man or beast; the bones in the blue clay cliffs of the Miocene are bones only, the flesh that clothed them is forever stripped away. But the life substance that actuated the one-time owners of these ancient relics is as potent today as it was forty millions of years ago when they were first laid down and it has continued through the centuries in uninterrupted flow. The stream may have altered course from time to time, may have swelled with figurative flood or dwindled with allegorical drought, but it has never ceased. From life to life the mysterious quantity has been passed along equally to all forms. Each individual has taken the gift momentarily. Whatever the type, high or low, the life substance does not differ in kind, only in degree.

CHAPTER 4

THE GHOST WORLD

ON A SWELTERING DAY IN AUgust years ago when I was a boy I very nearly drowned. I had been walking in the Virginia pine woods all morning and when I emerged from the trees onto the beach I was hot and tired; the thought of a plunge in the cool Bay water was enticing and I stripped off my clothes and waded in. With leisurely strokes I swam out about a hundred yards to a place where some ancient pilings protruded from the water. Once there had been a steamboat wharf at this point but it had long since crumbled and only a few old timbers still hung together. When I reached the place I climbed up on one of the few remaining planks and wearily lay down and went sound asleep.

When I awoke about an hour later, I noticed that the tide had turned and was flowing strongly in. Being in no hurry, I idled a little longer and then slipped into the water and swam for shore. The water was fairly deep and did not shoal until a few feet from the beach. But I gave the matter no thought and slowly paddled along. Suddenly my arm began to burn as though on fire and as I instinctively recoiled I saw that a number of white stringy tentacles were drapped over the flesh. I had blundered into a jellyfish.

This was a common enough occurrence, for these creatures occur in the Chesapeake in numbers and almost everyone who has gone swimming in the Bay has, at one time or another, been stung. A jellyfish sting is accepted by the people who frequent the Bay county as an unexciting, if unpleasant, experience. So, after rubbing the injured member for a minute or two and blaspheming a little, I continued. But I had hardly taken four or five strokes when a leg began to smart also, and then the other. Irritably, I looked out to free myself from the jellies and then with head lowered for speed, swam squarely into the fully-extended tentacles of another. For a brief instant I could feel the flabby touch of its tissues, then a wave of pain swept my face as though someone had poured strong acid over it. One or two tentacles dragged across my shoulders, leaving long red welts.

With lashing hands I drove away the creatures, rose to the surface, and with face aflame opened my eyes. To my dismay I saw that I was surrounded with the close-packed bodies of dozens of jellies. Somehow, by some trick of the currents, these jellies had all been collected in one swarm and were floating in on the tide. Behind their steadily pulsing umbrellas hung long streamers of whitish poison-laden tentacles.

Turning I looked toward the pilings. The medusae were thick in that direction too. There was no help but to swim through them to the shore. As far as I could see the water was dotted with their bodies. For once, in all my wanderings about the Bay, I was really disturbed. I already felt on fire and I dreaded to touch any more of the burning tentacles. While I was treading water wondering what to do, another one drifted against my flesh. The sudden irritation settled the issue and I struck out, swimming as hard as I could go. At almost every stroke I ran into a jelly and the pain became maddening. I have spent years about the water in all sorts of circumstances but I never came so close to panic as then. Long before I neared the beach I felt that I could stand the torture no longer and there was a terrible minute when I felt myself losing control. My chest, limbs, stomach, back, arms, and face burned frightfully; because of the pain I forgot to swim, began gulping water, and felt myself going down.

I have no idea how far I went nor how long I was underwater but at the last moment reason again gained the ascendancy, and I frantically clawed my way to the surface. Gasping for air, with aching arms I struggled for the shore. It seemed miles away but finally my toes touched sand and with a rush I pulled myself out and fell on the beach. Choking and coughing, I managed to rid my lungs of most of the water but there was no alleviation of the pain. From head to foot my flesh was streaked with bright red welts. For nearly an hour I paced up and down the beach, then slowly the pain began to subside. But for the remainder of the day I was hot and feverish and not quite back to normal until the next morning.

This experience was most unusual. It is only on occasion that the jellyfish swarm in such thick schools. Normally, they come during the height of the season in large numbers but they are scattered and rarely offer any serious hazard although people have been temporarily blinded by swimming into them with open eyes. Often a period of years pass and few stinging "nettles" be seen; in other seasons they appear by the thousands, floating everywhere. The reasons for their relative scarcity and abundance are not too well known.

For some time my unfortunate adventure curbed any interest I might have had in jellyfish, and I failed to appreciate fully what interesting creatures they really are. More than any other beings, they are truly animals of the sea, their frail, diaphanous shapes symbolize the life of the ocean in a way achieved by no other form. Like pale-colored ghosts they roam the waters of the world from the frozen arctic to the warm currents of the tropics and from the shallow bays and creeks to the uttermost regions of the great deeps.

There are a number of species and types which frequent the Chesapeake and they vary in shape and size from tiny transparent spherelike creatures to massive medusae with spreading bodies a foot in diameter and long trailing tentacles six feet or more in length. Probably the Chesapeake Bay, during the summer at least, contains as many jellyfish per square mile as any similar body of water in the world.

Their numbers are incredible. Once during a summer at Solomons, Maryland, I spent a number of hours beneath the surface of the Patuxent River near the Chesapeake Biological Laboratory in a steel cylinder, watching the life floating by on the tide. For a long time nothing passed the glass window except the transparent bodies of thousands of jellyfish. They were so filmy and so blended with the water that at first I was hardly aware of them.

Then, as the minutes ticked away, I began to realize the enormous mass of almost invisible life that was surging by. I began to keep count. The creatures were ctenophores, wraithlike comb-jellies belonging to the genus *Mnemiopsis*. For convenience I marked off an arbitrary space of six square feet and counted only the ctenophores that passed through this area. For several hours I tabulated jellies and discovered that an average of forty-eight went past each minute; this totaled 17,280 individuals of one species alone for the entire period. The Patuxent River at this point is about a mile wide. By computing the cross section of the river and making the necessary calculations I estimated that no less than 1,218,816,000 comb-jellies passed up the Patuxent during that single tide! Consider that the Patuxent is only one of several dozen small rivers in the Chesapeake region and that this same thing was going on all over the Bay from the Port of Baltimore to the Virginia Capes nearly seventy miles away. The total number must be astronomical.

The most familiar jelly of the Bay and the most despised is the famous *Dactylometra*. It is the bane of swimmers and the same creature which nearly caused my demise as a youngster. The hate with which it is usually regarded obscures the fact that it is really a graceful and even beautiful creature. Its umbrella is an almost hemispherical body edged with forty-eight little lappets like the scallops on old-fashioned lace curtains. Long tentacles trail out behind in a graceful train. Between them and suspended from the mouth, which is beneath the umbrella, are four long crenulated veillike mouth arms.

The colors are variable; one form is milky white, others are yellowish, some even bluish; the most spectacular of all have blood-red rays radiating from the center of the umbrella much like the symbol of the rising sun seen on the Japanese flag.

Dactylometra can be classed as among the most dangerous of jellyfish and its sting approaches that of the famous Portuguese Man-of-War of the tropics. I have been stung by both and there is little choice. The Portuguese Men-of-War are probably considered worse because of their large size; sometimes they have tentacles as long as forty-five feet. *Dactylometra,* happily, seldom exceeds five or six. Severe illness, even death, has resulted from jellyfish stings; the injury may vary from a slight burning sensation at the place of contact to generalized pain, high fever, and in severe cases prostration coupled with difficulty of breathing.

Not all the jellies are stinging. The comb jelly, *Mnemiopsis,* which occurs in the Bay in such untold numbers, lacks tentacles harmful to man and it may be handled with impunity. Indeed, it is almost impossible to go swimming in the Bay in summertime without bumping into their frail forms. *Mnemiopsis* are roughly pear shaped and are quite transparent. From the surface they are hard to see, but observed in their own element they are often beautifully iridescent. The iridescence is given off by their bands of waving cilia. These are minute hairs arranged about the body in orderly rows like the teeth of a comb; by waving these hairs in rapid succession the creature attains a slow progression. While one watches, waves of green, pink, yellow, and blue light pass over the cilia and disappear only to be instantly renewed. The effect is one of utmost delicacy.

There is another comb jelly which frequents the Chesapeake, particularly in the fall of the year. This is *Beroë*. It is as frail as *Mnemiopsis* but, unlike that jelly, it is not colorless. Instead, it is tinted with a shimmering lavender and pink. At times this assumes a brilliance and possesses a sheen that has to be seen to be appreciated.

Beroë is peculiarly shaped, usually being somewhat flattened in the form of a square shield. Like all comb jellies, they are helpless plankton and are carried wherever the current takes them. Except for a slight directional movement created by the action of the cilia, they yield themselves to the water and to whatever fortune tide and circumstance provide.

Not the least interesting of the Bay's medusae is *Aurelia*, the moon jelly. It is found only in the lower Bay although a few individuals are seen at times as far north as the Patuxent River in Maryland. As the name suggests, it is pale silver in color and attains the size of a dinner plate. Its tentacles are short and there are no trailing appendages. When viewed full face, so to speak, it resembles nothing quite so much as a moon which oddly and inexplicably has somehow become misplaced and fallen into the water. Its movements are slow and deliberate.

The *Cyanea* jellies also are common at times. Like the *Dactylometra*, they sting but not so badly. They are given to beautiful pinks and lavenders and possess about eight hundred delicately colored tentacles arranged in eight clusters about the umbrella. These organs are highly contractile and may be withdrawn out of harm's way or expanded until they are about twenty-five times the width of the animal's body. The *Cyanea* jellies are among the largest of known jellyfish. Individuals have been seen measuring six feet across the umbrella and carrying tentacles 120 feet long. Jellies of these dimensions can undoubtedly kill large fish, and any animal which becomes enmeshed in the tentacles of such a monster must suffer a sudden and horrible death. Fortunately, *Cyanea* of these dimensions are rare; those which inhabit the Bay are relatively small, being at most four or five feet long when fully extended. The big ones live in the open ocean.

There are numerous other Chesapeake jellyfish. Most of these are so small and transparent that they are seldom noticed; others are of good size. All are strange, bizarre, and exotic looking.

With this brief recitation I have introduced the subject of jellyfish and have sketchily pictured the more common Chesapeake forms. But at this point the pen falters and I find myself devoid of adequate simile to describe further these fantastic animals. For they are exactly that. Of all the creatures of the earth, except possibly the octopi, the jellyfish are the least reasonable of beings; their structure, their mode of life, their means of genesis, their very existence is a source of wonderment.

Over ninety-five per cent of their entire bulk is water. This is literally true because I have weighed their bodies newly removed from the Bay and then weighed them again after placing them in the hot sun to dry. They shrivel to the merest film of tissue—so thin and fragile that it crumbles at the touch. Their body moisture is usually like that of their surroundings but in some few forms, particularly the *Aurelia* and *Cyanea,* the moisture contains more potassium and less magnesium and sodium than that of a present-day sea water. It has been suggested that their salt content represents that of a prehistoric sea in which their ancestors some million generations removed lived and roamed. And lest this be put down as mere rhetorical fantasy, it should be remembered that the ratio and type of salt in human blood very nearly approaches that of the ocean and that our ears are really modified gill clefts.

Yet, for all their watery substance, these oceanic beings fulfill all the functions required to maintain themselves and to reproduce their kind. Although so fragile that they cannot be handled out of their natural element without damage, they manage to capture and devour creatures many times more powerful and solid than themselves. Fast-swimming fishes and armored creatures are victims of these seemingly helpless jellies. Even those types which are devoid of tentacles capture quantities of food, and so efficient are they that the passing of a large school of jellyfish has been known to depopulate an area of its previously abundant floating small life.

In their manner of securing food the jellies and their near relatives, the hydroids and anemones, utilize a system which is remarkable and intricate in its mechanisms. The majority, of course, kill their prey by stinging or paralyzing their victims so that they may be drawn into the mouth and devoured while still alive. The marvel, however, is not so much in the fact that the tentacles or other organs sting, as in the technique by which they function. If one of these long fragile appendages is examined under a strong lens it will be seen to be composed of a host of little cells usually appearing as oval or pear-shaped capsules. These capsules contain minute tubes usually coiled tightly like microscopic watch springs. The tubes, however, are not springs at all but deadly poisonous arrows.

Patiently, the darts lie in wait until the cell is touched by some object. Then with lightning rapidity the waiting springs uncoil and shoot into the flesh. Once embedded, the paralyzing poison is squeezed through the hollow tubes to do its deadly work. They are the world's smallest hypodermic syringes.

The darts themselves are objects of some complexity. In addition to being hollow they are often provided with numbers of wicked-looking spines and bristles. These are usually arranged in spirals running clockwise, and they ensure that the arrow will remain in the flesh until its function is performed. Certain jellies are provided with stinging devices which operate on a different principle. These do not pierce the flesh but are so arranged that when touched they lash out and, like a whip, coil tightly around any object which comes in contact with them. The barbs and bristles make the prize secure. This arrangement is highly valuable in the capture of such minute organisms as certain of the filamentous and lacelike larval crustaceans.

The poison itself has been extracted from the ground up tissue of tentacles. It is deadly stuff. Injected into small animals it causes paralysis, failure of the respiratory organs, and death. In certain of its varieties its effects are like those of curari, the substance used by the South American Indians to poison their arrows and which dulls the ends of the nerves, deadening them until they no longer function.

Those jellies, like the ctenophore, *Mnemiopsis*, which have no long trailing tentacles, utilize still another system for securing their prey. These animals feed on the small beings that live by the billions in the Bay water. The larvae of crabs and oysters, copepods, fish eggs, newly hatched fishes, worms, and other microscopic creatures all fall prey to their appetites. As these come close they are pulled into the body by numerous cilia waving in unison. There they are trapped, like flies on molasses, by sheets of sticky mucus which flows into the mouth in a steady stream. Once entrapped, few of the creatures escape.

Several years ago in July, during one of those seasons when the *Dactylometra* jellies were abundant, I was standing on the Bay bottom in about twenty-five feet of water. It was high noon and a heavy wind was blowing on the surface; whitecaps were breaking and spreading their froth in wide streamers. The boat from which I had descended was jumping crazily up and down but near the Bay floor there was no hint of the turmoil except for the light captured by the facets of the waves and focused downward in long yellow shafts. Because of the disturbance at the surface, the jellies had all dropped some distance and one by one were gliding along on the tide. The diffused sunlight momentarily catching their fragile tissues high-lighted them and caused them to be accentuated against the dark-green background.

This particular day I did not mind them, for I was clad in a heavy canvas coverall and was safe from their tentacles. Thus I was able to devote my entire attention to their passing without fear of being stung. Much to my surprise about every twentieth jelly was accompanied by two or three *Peprilus*, little, silvery, compressed butterfish. One jelly came into view with five individuals clustered beneath its body. As the jellies swirled by, the fish, alarmed at my strange shape, threaded their way between the hanging tentacles, and took positions near the crenulated mouth arms and close under the umbrella. I almost winced as they went between the tentacles but they passed unharmed.

This association is one of the strangest phenomena of the undersea. It is not restricted to *Peprilus* although, to my knowledge, this species is the only Chesapeake form which deliberately keeps company with the stinging jellies; there are other fish, notably *Nomeus,* which are associated with the Portuguese Men-of-War and are seldom seen apart from this jelly. Imagine living in a space hung with dozens of high-voltage electric wires with just enough distance to slip between. Yet this is comparable to the everyday existence of the butterfish that keep company with their dangerous hosts.

It is thought that the purpose of this companionship is to provide protection for the fish against its enemies, and additional benefits may be derived in the form of particles of food lost by the jellies. It is largely a one-sided relationship, for it is most certain that the jellies are not aware of their fellow travelers. However, sooner or later a butterfish makes the fatal error of touching one of the waiting tentacles, and the debt of its tribe is paid in part. Their small half-digested bodies have been seen inside the stomach cavities of the medusae, slowly being absorbed. When the assimilation is complete, the jelly releases the bones and other unwanted fragments and these fall gently to the Bay floor.

How this association came into being is one of the mysteries. Perhaps a superbutterfish a thousand generations removed pressed hard by some hungry carnivore first discovered the protective value of the death-laden tentacles and in some unknown way passed the new-gained knowledge on to his brethren and to his successors. More reasonably, the association probably was a gradual development created by some need about which we know nothing; in time a casual association became habit, then an instinct common to all butterfish.

Jellyfish are most strange and unusual in their methods of reproduction. They may be sexual or nonsexual, male or female, or both in one individual depending upon the type. Some forms scatter their eggs through the water; others retain them until they hatch. The eggs may become free-swimming larvae, tiny transparent beings which dash frantically about, and eventually develop into free-swimming jellyfish more or less like their parents, or with equal facility they may attach themselves to some solid object and become so unlike their ancestors as to be unrecognizable as near relatives. These attached larvae grow into organisms which resemble nothing so much as aberrant plants. They have stalks, branches, and even buds. The buds may grow to resemble queer animated flowers; but do not be deceived, they are neither buds nor flowers. Yet, plant-like these pseudo-floral beings can send out shoots and stolons and develop other flowerlike individuals much as real plants do.

These individuals make free-swimming jellyfish by casting off the buds or other portions of their bodies. If a sunflower should inexplicably detach itself from its stalk and go suddenly floating about through the air by waving its petals, and then later drop its seeds to grow new plants, the simile would not be inept. The plant-animals are termed hydroids and in those species which utilize this form of reproduction a question arises as to their exact status. Whether the free-swimming jellyfish is merely a breeding form for the convenience of the fixed, immobile hydroid or whether the hydroid is only the young of the floating jelly is a moot point. A hen from the egg's outlook is only an egg's way of making another egg.

It is impossible to generalize about the birth of jellyfish. There are too few rules and too many exceptions. As a group they are consistently inconsistent; the normal is to be abnormal; the usual, unusual.

Consider the fantastic history of the Bay's best known jellyfish, the tentacled stinging *Dactylometra.* When the male reaches its maturity it pours forth into the open water a host of sperm products. These are distributed in prodigal quantity and undoubtedly countless numbers must be wasted, lost in the swirling currents or devoured by the eaters of tiny beings. But always a certain number find their way to other *Dactylometra* and there fertilize the waiting or newly released eggs.

In time, the eggs develop and form a tiny hollow organism which sprouts a growth of waving hairy cilia. These act as oars and scull the creatures along at a furious rate. It seems to have difficulty making up whatever serves it for a mind, for it dashes hurriedly in one direction, only a moment later to stop and scurry equally as rapidly in another. There seems to be no sense whatever to its antics. In a few days, however, the creature tires of its frenzy and settles on some solid object where it changes shape and comes to resemble a small bottle or, better yet, a tenpin. A sticky substance is then secreted which forms a sort of disk about the animal's base and anchors it firmly in place.

Then, having attained a sedentary life, the creature prepares for further developments. At the upper end of its body it suddenly produces a cluster of waving tentacles and assures itself a supply of food. The animal is now all of a quarter inch in height and about one-sixteenth of an inch wide. In this diminutive size it is able to produce other tenpinlike animals or, indeed, to create fully formed jellyfish if it so wills. If, however, it decides to produce its own shape it may do so by slowly moving away, leaving behind a minute disk which, in time, grows into another individual. This may go on indefinitely; new individuals mark the trail of the parental peregrinations. Imagine having a carbon copy of oneself spring up in every footprint!

Now, if any of these individuals wish, instead, to produce swimming jellyfish they will begin to show a number of ringlike constructions about their bodies, which become deeper and deeper until the animal appears like a series of saucers piled one on top of the other.

These saucers, unlike any known crockery, twitch and squirm as though unhappy at being sandwiched between their flattened brethren. Occasionally they go into temporary fits of pulsations, at other times they lie quiet and unmoving. As time progresses, the saucers become more restless and more frequently agitated as the constriction advances until in the activity of their movement the last thread of their attachment is broken and they are free to go where they will.

But not yet are these saucer-beings true jellyfish. Their bodies are not yet round, nor are they equipped with the crenulated trailing mouth arms and long tentacles. Instead, they are somewhat starfish shaped, but with four arms instead of five; unlike starfish, however, they are quite filmy and transparent. And they are very small; at their fullest expanse they do not exceed three-eighths of an inch.

But in a short while, unless they are devoured by their cannibalistic relatives or meet some other unhappy fate, they fill in the gaps between their arms, develop little fibrous tentacles, and become miniature medusae. At this stage they are still small, about a half inch in diameter; but before long, if the hunting is good, they grow, increasing steadily in size until they attain full maturity.

Perhaps the strangest and most entrancing sight I have ever seen in all my trips beneath the waters of the Chesapeake occurred one evening some years ago when I descended from the deck of a fishing boat in the lower Bay near the mouth of the Rappahannock River. It was one of those calm hot nights such as often occur on the Bay in July and August. Under a full brilliant moon the Chesapeake lay still and calm and quiet. Only the distant stirrings of the night birds on shore, the drone of a few wandering mosquitoes disturbed the hush. Beneath the surface the light of the moon stole through the water with a soft glow, a pale radiance more suggested than seen. The olive color of the day had disappeared; in its place was only black and gray toward the depths, the lunar luminescence above. I stopped in midwater and gazed upward. As I watched, three large medusae in succession passed silently between me and the surface. As they drifted by, the light caught their translucent tissues and burnished them with glowing silver. The effect was startling. Against the dark water they were cast in brilliant relief; the light accentuated their every feature. Slowly, gently they expanded their umbrellas and pulsed ahead, throbbing on the tide that was flowing toward the open sea. The silver faded to pale ivory, then pearl gray. For a few seconds they remained in view and then, as quietly as they had come, they melted into the distance and were gone.

CHAPTER 5

THE CHESAPEAKE MARSHES

NEAR THE TOBACCO MARKET town of Upper Marlboro in Maryland, the Patuxent River finds its way to the level of the sea. After a long tortuous journey from its source at the edge of the Piedmont Plateau and across the tumbled region of the Eastern Fall Line, it suddenly emerges from a narrow green valley overhung with trees into the broad reaches of the Patuxent marshes. Here it changes from a swift, tumbling stream, with rapids, eddies, and even waterfalls, to a slow, winding, lazy waterway. The change is complete, and as far as Holland Ferry, twenty miles below, the river is unlike either of its extremities.

These twenty miles of river constitute a sort of halfway zone, a place that is neither part of the open salty bay nor a portion of the upland country. The upper end of the swamp is steeped in the fresh water of the flowing river; farther down the water becomes more saline as it merges with and gradually becomes part of the Chesapeake.

In a sense, this lovely valley of wild rice and waving cat-tails, this midway region of water lilies and marsh grass, of narrow channels and mud flats is a place of struggle, although it is a strife which goes on so quietly and so peacefully as to be almost unnoticed. It is the scene of the never-ending battle between the ocean and the solid land.

Twice in a day the long fingers of the Bay creep in between the reeds and the cat-tails, flood the spaces between the grass hummocks, and fill the curving channels with a brackish inundation; twice the land regains what it has lost. As the ocean retreats it takes with it particles of mud and sand, bits of leaves, fragments of floating reeds, blades of broken grass. The land in its turn sends out roots and tendrils to anchor the soil more firmly; seeds fall on the bars exposed at low tide and secure themselves, grow, and extend their hold. The river sends down new ground; this falls to the bottom in the slack tide and creates new soil for green growing things. Everywhere, on a small or large scale, the struggle continues. The land fills up an old channel with tightly held earth; the waters cut a new one, etching out the grains one at a time or in big chunks, dragging them out to sea.

Similar strife is being waged all over the Bay country,

wherever the sea meets the land. On the ocean front the battle is thunderous and is waged with big waves and giant combers; in the valley of the Patuxent and in other marshes of the Bay, the action is slow and stealthy, silent and furtive.

This alternate expansion and recession, this twice-daily drowning of the marshes and their drying out again sets the tempo for the life of the swamp and limits and controls the growth and character of the vegetation and of the swamp itself. Plants and animals, mammals, birds, fish, and serpents all lead their lives in accordance with the periodic invasion and retreat. Consciously or unconsciously, all the swamp beings fall into the pattern of the tides and become a part of the earth-sea struggle.

The Patuxent marshes are not greatly different from most of the Chesapeake marshes except that they are larger than many, and longer and more winding. The tidal marshes are a universal feature of the Chesapeake and those of the Patuxent are typical. Almost every estuary has its bordering swamp and the shores of the Bay are dotted with thousands of lagoon marshes. These are places varying from little patches, hardly a half acre in extent, to large areas covering miles of half-submerged land. They are formed, usually, when the currents create a sand bar or shallow ridge across an indentation in the shore line. This bar, building itself ever higher, soon affords protection for the shallow, still water inside. In time, in a few days, a few weeks, or months, seeds lodge in the exposed flats, take root, and anchor the soil. Leaves and branches fall from nearby trees; rain-washed silt from the neighboring land fills the interstices with soft, silky mud; plants grow, spread their leaves, and blossom only to crumple in the water and thus add their bulk to the total. In this way swamps are born.

These marginal swamps are places of surpassing beauty, and there is a quality of loveliness about them that is different from all else. Even when they occur in regions that are thickly settled, they are set apart in lonely fashion, protected from the world by their dense fringing vegetation or by their acres of yielding soil. In a world of busy, swarming human beings and whirring machinery they remain isolated and free of intrusion. It is, perhaps, this unspoiled character and their untrammeled natural appearance that make them most appealing. There is also a brooding air that is most entrancing, a vague somnolence associated with few other scenes.

The life of the swamps is one of cycles, of growth and recession, advance and retreat. This is not only daily, tide urged, and controlled, a sort of natural breathing, so to speak, an oft-reiterated systole and diastole of movement, a falling and rising of the waters; it is also the alternation of day and night, the waking and sleeping of the creatures of the swamp, and the evidence of their activities. It is the swing of the seasons, the turn of spring to summer—fall to winter—the coming of the heat and of the cold, the burgeoning of the seemingly dead seeds and roots, and their return to brown lifeless substance again. It is the coming and going of the ducks and geese, the sandpipers, and the marsh wrens; it is the hum of insects and their later silence; the difference between the shimmer of heat waves above the green reeds and the crackling of ice between the broken brown fronds.

It is not possible to describe a Chesapeake swamp in one unhalting paragraph, nor in two or in two times two. My personal recollections are made of a host of separate and isolated remembrances, like the numerous pictures in an old-fashioned art gallery or the facets of a many-sided crystal. A Chesapeake marsh is a multitude of mind portraits done in a host of colors, but always with delicacy.

It is the sight of a white egret posed on one leg against a background of green reeds, waiting motionless and patient for a fish to pass; it is the opened chalice of a great white and purple, yellow-stamened mallow, or the equally beautiful vision of a water lily in full bloom beside oval lilypads floating on dark water. It is the somber green of overhanging pine trees reflected in the still pools; the widening and often-repeated circles of raindrops falling on the winding channels, making a soft hissing sound as they touch. It is the splash of fish and the liquid burble of a marsh wren rising in a burst of energy above all other sounds. It is sunshine and blue sky and sparkling waters; it is also grayness and dull brown; ice and a piercing wind; it is the noise of dried reeds clattering one against another, millions of minute rubbings and bumpings blending into one multitude of sound; it is the noise of ducks and geese gabbling to one another in the early morning; the sporadic clattering of rails; and the rustle of some unseen being creeping between the cattails. It is the gleam of moonlight upon a layer of mist lying close to the water, and the reflection of starlight; conversely, it is glare and heat of quivering hot air. It is the smell of dried mud and of decaying vegetation inextricably mixed with the aromatic scent of pine and the odor of long-dead fish and dried mussels. It is an expanse of waving grass turning golden and swaying and bending like wheat in the breeze, rippling and waving; it is the red glow of the setting sun and the faint clicking of bats in the long shadows; it is the dancing of swarms of hovering gnats and the drone and whine of mosquitoes in the gloom. It is the red, black, and yellow bodies of painted turtles sunning themselves on half-submerged logs; the V-shaped ripples that denote the heads of serpents gliding across a channel to seek frogs on the other side; frogs themselves calling in the dark, the shrill of toads, the deeper, solitary tone of the larger batrachians, and the massed chorus of thousands of spring peepers. It is the clustering of a glistening mass of dewdrops on a strand of marsh grass, and the patterned cracks of mud dried in the sun. All these things and many more, visual, sensory and aural, are a part of the mosaic that comprises the total picture of a Chesapeake swamp.

And so the portrait of a marsh is not an assemblage of water and soil and vegetation which may be coldly defined and catalogued. Like music, which is conceived of notes and scales, of chords and arpeggios, it is a never-ending series of compositions; of Bach and Beethoven, Strauss and Debussy, of Chopin and even Wagner. It is a processional of portraits, a series of dramas, some small, some large, a sequence of intransient masterpieces on a variety of scales; some are grandiose and majestic like the murals of Michelangelo; others are delicately and gently limned like the threads of a Chinese silk or the tracery of a Dresden porcelain.

Once I spent a whole twenty-four hours on a bluff overlooking the Patuxent River in the very heart of the swamp. In a canoe I had found the spot in the evening and had made camp among the pines and had spread my sleeping bag on the brown needles. Tired from a long paddle I decided to remain until the following day. The place was well chosen and quite isolated. From the height the full extent of the river valley was visible. The masses of reeds and wild rice and the winding course of the river itself, curving back and forth like some great glistening serpent, were spread beneath my feet; from the pines I could see but was unseen and unnoticed.

I awoke about three in the morning. It was dark and cool. There was no wind and a deep hush had spread over the world. Silently I pulled on a sweater and a pair of moccasins and softly, as though half afraid to mar the quiet, stepped over to the edge of the bluff and sat down with my back against a gnarled old pine. There was no moon and the marsh was dim and obscure. Very faintly, scattered stars peeped out between masses of dark clouds. Here and there a few caught the surface of the water and gleamed palely in reflection.

For a long while I listened and heard no sound. The night creatures, tired from the evening's activity, had ceased their calling and wanderings and had retreated into their hiding places, or drowsed invisible in the gloom. It was their hour of waiting before the change from night to day, when all the world was in suspense.

I must have drowsed because when I woke again I was shivering and a heavy dew saturated the grass and dripped from the leaves. It was still dark but the darkness was not quite so black as before. The swamp was still invisible and was still silent. But as I listened there came from far out in the center one low, clear note. For a long time it was the only sound, then I heard it again, stronger and more full. In the daylight, or at any other time, it would have passed unnoticed, drowned in a medley of other sounds, but in the night it arose alone and unnamed, and because of its loneliness dominated the dark.

The note reached its full pitch and then softly died away. Somehow I was reminded of a rocket I once saw at sea, a single pinpoint of light that soared up out of the blackness, blazed into brilliant illumination at its zenith, and then vanished from the earth. This single note was the signal for which the marsh had been waiting. Once again it came; then far off in the distance another pierced the gloom, then another. The spell was broken. In a steady progression other voices added themselves to the first until the air was throbbing. And all the while the night retreated, giving way to gray-black, to gray, and then turned soft rose.

With the beginning of light the whole swamp came alive. There was the rustling of thousands of wings as the hungry flocks began to feed and to go about their early errands; unseen splashes betrayed the activity of small fishes schooling, and the querulous voices of hidden ducks echoed back and forth across the valley.

The amount of life that is hidden and becomes evident only in these early-morning hours is surprising. The sunlight had hardly touched the tops of the distant trees when large groups of red-winged blackbirds were wheeling and streaming back and forth from their favorite feeding places; flocks of sandpipers coasted just above the mud flats, then settled at the water's edge and tripped over the sand bars looking for food exposed by the low tide. Like brown shadows the henlike bodies of rails could be seen threading between the reeds, occasionally one punctuated the morning air with its clatter; coots paddled in small groups of two or three along the river's brim; several flocks of crows crossed the valley cawing as they went; one of these suddenly turned and dived out of sight along the distant bank; hosts of diminutive sparrows swayed on the tops of old reed stalks or undertook short trips between the swamp and the shore; myrtle warblers clustered about the bushes and low shrubs or swirled in miniature flights between points of land, descending on the bordering trees like leaves blown by an October wind.

The sequence of events ushered in by the coming of the day followed a slow but definite pattern. The first joyous activity caused by the departure of night reached its climax as the rays of the sun passed the distant treetops and then stole downward, catching first the very tips of the reeds, leaving the bases in shadow. At this moment every feature of the swamp was accentuated, cast into vivid high light. Every reed carried momentarily a crest of flame; and the dew that saturated every fiber caught the color and cast it back with a brilliance reminiscent of the shimmer of old silk.

With the advent of full daylight the activity slackened; the flocks still moved, though in smaller numbers and with less vivacity. As the dew disappeared and the heat of the sun began to temper the morning cool, the voices abated, diminishing steadily, almost imperceptibly. The fish no longer jumped as merrily and their splashings were less frequent. The sand bars and the bobbing figures of the sandpipers began to disappear one by one as the rising tide stole their feeding places and crowded them against the reeds.

By ten o'clock the scene had changed, the mood altered. The marsh was then drowned in sunlight and the thermometer had climbed; heat waves were beginning to rise. It was no longer the hour of the ducks and coots, of the red-winged blackbirds or the rails. They were nowhere to be seen, nor were their voices to be heard. Instead, a low steady drone rose from the marsh, a vibrant monotone that persisted minute after minute. It so permeated the air and it came into being so slowly and unobtrusively that at first I did not notice it, and I was unaware that it had not been there before. Only when the coarser sounds had been eliminated did it come to attention.

It was the sound of millions of insects, the drone of hundreds of bees, the hum of countless flies, the beating of the elytra of beetles, the stirring of all the legions of arthropods. There were individual sounds that rose slightly above the general hum as some insect passed close; the high-pitched whine of assassin flies; the peculiar rustle of dragonfly wings and the irritating singing of tiny, almost invisible diptera. But it was the bees that dominated the ensemble. One does not usually think of a salt marsh as filled with bees; they are more reminiscent of clover, timothy, and meadows of verdant grass. But they were everywhere, swarming about the just-opened mallows and other swamp flowers. Yet there must have been some other attractions, for the bees were many and the flowers few.

The hours of the insects were long and as the sun approached the zenith they persisted and their ceaseless monotone increased in volume until it overcame all other sounds.

By then the tide was high and the *Littorina* snails, those gray comical periwinkles which spend all their lives on grass stalks hovering between successive inundation and dehydration, were soaking up the brine and had emerged from their lime houses where they had been waiting tightly sealed for the coming of the water. Their life is one monotonous succession of wettings and dryings, of slow creeping movement while they are feeding and long periods of sleep between times, tightly closed to keep from drying out. They are truly midway creatures in a midway world; their lives are limited to within a foot or so of tide lines; in their slow methodical way they have partly broken from their ancestral ocean, yet have not quite achieved a life on dry land. They may stray just so far; beyond a certain point is death by dehydration. The full tide is their necessity, and although they no longer need to spend all their hours in the water they are held to it as though bound by prison chains.

The division of the day into the hours of the events of the swamp is a much more meaningful method of keeping time than our mathematical chronology. The Chinese have long used this poetical technique in the designation of the years and the seasons. To have lived in the year of the Dragon or of the Tiger is far more entrancing than to have existed through 1936 or 1943. It may be argued, of course, that such a method is unscientific, and so it undoubtedly is. But to attempt to describe the passing of a day in a Chesapeake marsh by cataloguing the arithmetical hours is an uninspiring business. How much more descriptive it is to speak of the Dark Hour of the First Voice, the Time of the Wakening of the Birds, the Interval of the Rising Tide, or the Hour of the Littorina! The creatures of the swamp do not live by arithmetic; they are moved by events, are actuated by sun and tide, by light and darkness, by heat and cold, by hunger and fullness, and by the movements of the life about them.

And so the time of the *Littorina* was succeeded by the hours of somnolence and of the falling tide. The sun passed the zenith and began its earthward trip. The heat increased and waves of hot air caused the distant shore and the reeds to quiver. The birds vanished, all except the vultures which endlessly soared high in the sky. Although they were the only moving, living things, they increased the feeling of drowsiness. For a long time the swamp lay still, quiet and deserted.

Such periods of calm always precede times of great activity. It is as though nature, asleep, is gathering force once again to show a burst of energy. The calm before a storm, the deceptive warmth of Indian summer, the slow gathering of the clouds before a northeaster are phenomena in kind. The heat of the day beats down the feathered things and the glare drives the mammals to shelter. Even the serpents seek the shade and the turtles retire to sheltered nooks. Only man is foolish and sweats and toils in the fields. But with the passing of the hot hours the burden is lifted and the world once again breaks into activity.

But there is no comparison with the alteration from dark to dawn. There is no time of the opening voice, no sudden ushering in of animate existence. Slowly as the heat declines and the rays of the sun grow longer, as the shadows steal in, the creatures of the place come out of their hiding and go about their errands. Serpents cross and recross the winding channels looking for food; ducks and grebes appear as by magic; the rails resume the clattering they halted in the early morning; the rush of wings is heard again; fish splash; the little green herons stalk daintily in the shallows spearing minnows. From the bordering pines mourning doves call plaintively. There is a different quality about these sounds and movements. The spontaneity is gone; the bird calls are superficially the same, but the vivacity is missing, a certain joyous timbre lacking. There is activity, but it is activity dictated by need alone; stomachs must be filled and preparation made for the coming of the night. Muscles are tired, the day has been long; there is a brief resurgence of energy but little is left over for joyous singing or carefree romping. The swallows are an exception. Their flashing bodies turn and twist over the marsh in endless gyration; their small-voiced twittering seems as sprightly as ever.

There is no sharp cleavage between day and night. The two worlds merge imperceptibly; the creatures of the day disappear singly or in flocks; their places are taken by the myriads of the evening. The alteration is not readily visible; even from my vantage point overlooking the swamp all the details were not revealed. As the shadows spread out and enveloped the earth, one became conscious that the red-winged blackbirds were no longer moving, that the mourning doves had ceased their interminable calling, that the sky was free of the last vulture, that the warblers were not to be seen nor were their songs to be heard. Only the *Littorina* remained impassive on the grass stalks waiting patiently for the coming of the tide. All over the marsh and in the bushes countless heads were being thrust under countless wings, countless pairs of eyes were closing, dropping into that half-sleep which characterizes the uneasy rest of the wild.

And, as imperceptibly as the day creatures disappeared, so did the night animals begin their activity. Seemingly out of nowhere, the twisting, fluttering shapes of bats appeared, replacing the swallows. Their darting forms silhouetted momentarily against the sky, then became invisible against the hovering shadows. The soft whisper of their tissue wings and the strange clicking they make sounded temporarily as they passed close, subsided, and then rose again.

I could not have told when the first frog voices made themselves heard. They were almost in full volume before I thought of them. In the nocturnal chorus the individual species could be readily distinguished; there were the shrill reiterated notes of the spring peepers, the deep "chung" of the great bulls, the protracted shrill of toads, and the similar but more musical song of the swamp tree frogs, the bright green, waxy-looking, big-eyed *Cinereus*. So gradually did the amphibians take up their chorus that it was not until the gloom was pierced by a frightened, almost childlike cry that I became aware of them. Somewhere in the maze of reeds a brown water snake had seized one of their number, and in quick terror the frog had cried out. Suddenly every frog voice in the whole vast extent of the swamp was stilled. For a long minute the silence persisted; then in the distance a frog called softly, then louder. Another took up the notes and soon they were in full chorus again. Frogs forget easily.

The sun had been gone only a short while when the whippoorwills started. The swamp echoed to their repeated pleas to "whip-poor-will." Every Chesapeake marsh has its quota of whippoorwills; they do not live in the swamp itself, but on the borders, preferably in the pines; and in the appropriate season, when they are calling, they set the tempo and timbre for all the marsh sounds. One seldom sees them, for they are habitants of the dark; during the day they remain quiet, motionless, sitting lengthwise on some log or branch, mottled or speckled, invisible against the wood. But as soon as night comes they begin their feverish calling, over and over, endlessly admonishing the world to "whip-poor-will," then again and again to "whip-poor-will," "whip-poor-will," "poor-will."

The time of the calling of the whippoorwills is a long one. They must derive a sheer joy out of it, some type of avian ecstasy that drives them on and on, or they are impelled by some vital, uncontrollable urge to communicate their message to their kind and to the wide, wide world. They are sort of a Temperate Zone "brain-fever" bird, and indeed their repeated calling drives some urban and unornithological souls nearly frantic.

The life of the early-evening hours is as busy as the corresponding burst of activity in the early morning. This is the time of the muskrats, the period when they go about the canals in search of food or sit on top of their feeding platforms or on their conical reed houses and survey their darkened domain. It is also the hour when the owls are soaring over the swamp looking for those same muskrats and for their lesser cousins, the mice. These latter beings occur in legions and their down-lined nesting places are found wherever there is a dry space.

The doings in a Chesapeake marsh at night can only be inferred; our eyes are too ill designed to give the needed vision. It is as well, for the twin senses of sound and smell are neglected human attributes. The call of the whippoorwill is more potent than the sight of one; the wonderful smell of dried mud, sun-baked reeds, decaying vegetation and the ever-present aura of salt water that drifts out of the marshes carries as many connotations as visual appearance. So, from my observation place upon the bluff, I could picture the events of the night without seeing.

The sudden slap of some object upon the water followed by a splash would be the hurried retreat of a muskrat aware of danger, perhaps in the form of a hovering owl. All over the swamp similar slaps and splashes indicated that the warning had been heard and was being passed along. Out in the open water between the grass islands a rapidly reiterated pattering sound accompanied by a pulsating whirring marked the place where some duck or other wild fowl was blundering into the night, literally walking on the top of the water, trying to take to the air. Something had startled it, and it was frightened, was seeking escape from its enemy, real or imagined.

Most of the night sounds are soft and must be listened for; they occur as undertones to the din of the frogs and the heavier sounds of the swamp. There are multitudes of tiny scrapings and raspings, little scratching noises which are difficult of definition, the sound of dripping, and the barely audible "plop" of bubbles of marsh gas rising to the surface and bursting. Back of all these is a faint clattering, an infinitesimal banging and bumping, in tune and associated with a delicate sighing. At first this is indefinable but then, as the categories of sound separate and take their proper places, it becomes plain. It is the sound of the night wind; the sighing caused by the passage of air through millions of grass stems, between the close-packed blades; the clattering and banging is the hitting of the individual reeds and stems, as they nod and bend to the pressure.

Slowly the hours of the night wore on. At some ill-defined point the whippoorwills ceased their calling. The frogs sounded less enthusiastic and presently the reeds and cattails were quiet again. Even the splashings and patterings were heard no more, for the wind had died, gently and gradually. Then came the most entrancing time of all.

From the top of a strong-scented myrtle bush close to a starlit pool a lone mockingbird began its song. Clear and sweet it poured out its whole repertoire to a silent world. Note after note went pealing across the quiet reeds, some soft, some low, some strong and full. What had wakened this dull gray bird and caused it to burst into song, there is no way of knowing. There was no answering call or another sound. Alone and for nearly an hour the melody continued and all the while the hush prevailed. Perhaps the very quiet, and the opportunity to be the sole actor in a midnight show caused this bird to break into melody. More likely, the song was created by some inner need for expression, some strange desire to fill the silent spaces with sound or to give vent to a sense of well being. Whatever the cause or whatever dreams prompted the song, it continued until long after the stars had reached the zenith and were dipping downward. Then on a last clear note it ended.

It was the final act of the day. The cycle had come to completion. For I drifted off to sleep and when I awoke, the dew was heavy on the grass, the morning chorus had begun.

CHAPTER 6

THE LANTERN BEARERS

ON A LATE SEPTEMBER EVEning the sun set in a great bank of black clouds. The clouds spread, and a premature twilight darkened the Chesapeake, turning the water to a somber gray. The day had been very hot and the air was heavy. From between the piled-up cumulus, sheet lightning flared fitfully and lit up the distant Virginia shore.

The Bay lay still and quiet and only a gentle heave coming in from the ocean, many miles away, disturbed the flatness of the water. The sails of our yawl hung slack and limp and the hull turned idly as it drifted on the tide.

For a long time there was no wind, nor any motion except for the turning of the ship and the expanding bulk of the clouds.

Then, when it was completely dark, when the stars had been blotted out and the only lights to be seen were the flash of the Wolf Trap lighthouse lying to the west and the faint gleam of the binnacle, a breeze came out of the dark. At first it was no more than a faint whispering, a slight stirring at the masthead, a warm breath against the cheek. It came again, stronger, and then steadily burgeoned into a full southwest wind. In half an hour the rigging was singing, the sails were taut and full, and the sheet lightning showed whitecaps speeding past in the dark. By midnight a half gale was blowing, although the air was still warm and the ship with the wind on the quarter went boiling up the Bay pushed on by both wind and waves.

Shortly after midnight the flare of the lighthouse at Smith Point, Virginia, began to show in the distance. It lay slightly off the port bow, as it should have; the course was fair and we set the ship to pass Point Lookout across the Potomac with a straight run for Point-No-Point beyond. Satisfied, the ship's company chatted, dozed on deck, or braced themselves against the lurch of the cross sea that was coming out of the black open space of the Potomac. Everyone was half asleep, except for the helmsman, when suddenly someone yelled "Breakers ahead!"

Instantly everyone snapped awake, and with a thundering of canvas the helmsman brought the vessel into the wind, heeling her far over. The sheets were hauled, and with spray pelting across the deck we beat into the dark. For the first time we realized how hard the wind was blowing; the rigging began to sing in high-pitched tones and the rails were buried in bubbly froth.

Hurriedly we checked our position and took bearings on the lighthouse again. The course was correct and according to our calculations the nearest shoal water was miles away. A heave of the lead verified this belief; we were in sixty feet of water. Once again the ship was set on her course and with wide-open eyes we peered into the dark. In a few minutes there came out of the blackness a faint glow and it seemed to be the white of breakers; in a long line the path could be seen extending in either direction. Another quick heave of the lead showed seventy feet; the water was becoming deeper, not shallower.

We saw that the water beyond the outermost line of palely glowing froth was brightly lighted and that other breakers lay beyond the first. Indeed, as we edged closer it became apparent that the whole region beyond was lit as though with some unreal supernatural light. The outline of the rushing waves was plainly visible against the black sky and as they broke the froth blazed forth with unnatural brilliance.

Then we realized what we were witnessing. We were entering a zone of highly phosphorescent water and the line that we at first thought was breakers on a beach or a shoal was really the outer limits of the region. The dividing line between the black, unlighted waters and the glowing zone was sharply delineated; within the space of two or three feet the change was complete. The line of division went snaking off into the distance, curving in and out in gentle arcs.

The brilliance of the lighted area was surprising. As we plunged over the line of separation the hull became visible with a greenish sheen, much like the glow that comes off the hands and dial of a luminescent watch. Even the ropes and rigging, particularly those close to the water, picked up the light and reflected it palely. It was ghostly, we felt like a phantom ship on a fiery ocean, a sort of Chesapeake version of the Flying Dutchman. Into the depths as far as we could see the Bay was ablaze with light. Every square inch of the water was glowing, every ripple and bubble was sharply delineated, etched in phosphorescence. For eerie effect I have never seen anything quite like it, and if, at some future date, I should be called upon to describe my ideas of the River Styx I shall not picture it as a black somber stream but as a replica of this September Chesapeake.

The dark night, the half gale, the breaking waves, and the turbulence of the heavy cross sea combined to produce the optimum effect. The violent agitation of the water, the bubbling of the froth, and the surge of the waves, all served to excite the billions of organisms that imparted the light, forced them to put forth the maximum brilliance.

For nearly five miles our course lay parallel to the dividing line and we diverged sufficiently to follow it for its entire length. There were times when one side of the ship was all alight; on the other not even the surface could be seen from the rail. Presumably the waters coming down out of the Potomac were relatively lifeless, were devoid of the micro-beings that gave out the phosphorescence. At first we thought it might be a matter of temperature but a thermometer dropped on either side of the line showed a divergence of only 1° F., an insignificant amount. Some subtle difference, of salinity, of chemical composition, or other unknown factors, caused life to exist in inestimable abundance on one side, inhibited it on the other.

As we sailed north, the dividing line curved slowly to the east and then suddenly swept off toward the Eastern Shore of Maryland and we were once again in a pitch-black sea. But we had not finished with unusual experiences that night, for around three in the morning the wind suddenly hauled and came blustering out of the northwest, no longer warm but sharp and cold. And with the change of wind the clouds tore to shreds and disappeared leaving a clear sky spangled with stars. Soon we were fighting for every inch, the spray began whipping over the decks and it stung as it hit. We donned sweaters and oilskins and shortened sail.

Then as we pushed on, dodging the spray and huddling behind the cabin to avoid the wind, the northern sky began to quiver and pulse. Streamers of pale-green light, like those we had seen in the water a few miles back, began passing in bands across the zenith, fading and then appearing again. It was the aurora borealis, and for the remainder of the night we were treated to one of the most magnificent displays we had ever seen. Great spokes radiated from the distant pole, wheeled across the sky and descended to the horizon again, alternate bands of yellow, green, and blue chased each other in silent progression, streamers and waves of luminescence deluged the heavens until the morning sun chased them away.

The two displays—one from the outermost regions of space, the other from the depths of the sea—were the most magnificent I have ever seen of either phenomenon. That they should have come on the same night was one of those strokes of coincidence that occur once in a lifetime. Peculiarly, we know little about either. The precise causes of the aurora are as veiled in mystery as the reasons for the nocturnal flashings of the micro-beings that inhabit the oceans. Each in its individual way was awe inspiring. The ghastly lighted sea, peopled with thousands of shining protozoans to the square inch, each fully alive and leading a separate existence, flashing their beacons for whatever obscure purpose their nature decreed, was as arresting as the mental picture of the surging across the heavens of cosmic rays or whatever other forms of interplanetary energy cause the sky to break into color and waves of light.

I recall one other remarkable display of phosphorescence. Unlike the first, this happened on a very still night in a little brackish creek leading to the Magothy River in Maryland. I was not on a boat at all but was seated on the bank beneath a spreading oak tree which overhung the water. It was very dark and while I sat, straining to see the outlines of the river, it began to rain, great large drops that pattered against the leaves and fell on the water with a hissing sound. Because I had no place to go I remained where I was and huddled against the trunk to keep as dry as possible.

For a minute it rained slowly, then the heavens opened and a deluge of water came down. To my surprise the dark river suddenly began to glow and for the five minutes the shower lasted the surface was bathed in fire. The agitation of the thousands of drops striking the surface caused the previously unlit microorganisms to blaze with color.

The shower departed as rapidly as it had come and for the sake of a better view I slid down to the water's edge. The river was dark again but in the black region beneath the tree the drops were still sliding off the leaves.

As each drop hit, the point of its impact burst into flame and then in widening circles the light spread out and faded away. For several yards the blackness was marked with flashes of light succeeded by spreading circles, sometimes by circles within circles as the miniature waves created by the impact chased each other into obscurity. Some of the circles were still visible when they were several inches in diameter. Reaching up I shook a limb and suddenly a host of flashes and circles appeared. The rings widened, met, and passed through each other, delineated in fire; for a second they were outlined and then the water became invisible again.

I took a glass jar from my pocket and emptied it of odds and ends, stepped down to the river and filled it. Its outline was plainly visible as the water rushed in and as I adjusted its lid a few pinpoints of light drooled down the sides and went out. Back home, I dried myself and then in a darkened room placed a few drops of this river water on a slide on a microscope. At first I saw nothing but when I tapped the slide sharply with a pencil I saw a few sudden flashes. One was in focus and for a brief second I saw that it proceeded from a spherical object with a tiny quivering tail. The light appeared to emanate from numerous spots on the surface of its body but it was lit so briefly that I could not be sure. Once again I tried to cause the beings to show their lights. I succeeded, but none was in focus.

Switching on the light I rocked the lens back and forth and in a second found the object of my search. It was, as I had expected, *Noctiluca,* one of the protozoans. Its name, appropriately, means "night light." In the two or three drops of water on the slide were nine or ten individuals as well as a number of other microorganisms which

I could not at the moment identify. The *Noctiluca* were rather pretty; through the lens they showed traces of iridescence and their tissues were limpid and clear. All were roughly hemispherical with a deep groove running partway around the body. The little taillike organs vibrated rapidly and caused them to move erratically through the liquid. In the light of the microscope there was no evidence of light organs nor indication of any reason for their existence.

The possession of luminescent organs in certain animals is a scientific mystery. In some, of course, as in certain deep-sea fishes, it is a necessity, a means of lighting the abysmal darkness, or of attracting prey, or of establishing recognition between individuals of the same species, of finding a mate. But in others, as in *Noctiluca,* it appears to be one of those useless extremes to which nature goes without visible purpose. Certain bearers of light organs are, paradoxically, totally blind. But the most senseless example of all seems to be the case of a deep-sea crustacean that carries its lights hidden far within its body, amid the intricacies of its branchial chambers!

The height of uselessness, however, appears to be achieved by the famous *Chaetopterus.* This animal is found all along the Atlantic coast from New England to the Carolinas, and in other parts of the world along the borders of the sea in warm estuaries. *Chaetopterus* lives, commonly, at about low-tide level in the mud and sand at the edge of marshes, and in local areas it is abundant. In the Chesapeake it is apparently confined to the extreme lower portion of the Bay near the ocean. It is highly phosphorescent and glows with a brilliance matched by few other animals. But its light is totally squandered; for *Chaetopterus* lives in a U-shaped tube buried deep in the mud and ooze of its native swamps and tide lines. One can walk over hundreds of burrows and not be aware of the gleaming light beneath. This seemingly senseless loss of good light is equaled in futility only by the possession of the vermiform appendix in man and by the fantastic nose lappets of male turkey gobblers.

Peculiarly, for all its waste of light, *Chaetopterus* is a highly efficient animal and it may be described as the only completely illuminated animated suction pump. If at some date I should be required to formulate a catalogue of the world's most amazing creatures, *Chaetopterus* would be high on the list. *Chaetopterus* is a worm; and like so many of the marine worms it is utterly unlike our common garden variety. It is a fantastic animal composed of a mass of successive lumps, fringes, segments, and lappets and it looks like something an old Chinese artist might have composed deep in his opium, a sort of vermian dragon. Actually, however, the strange appendages of the creature serve most useful purposes. One group of these are used as pistons for the underwater suction device and are so shaped as to fit exactly the walls of the underground tube utilized by the beast as a home. This tube, a U-shaped affair, is buried in the mud with either extremity protruding slightly above the bottom. It is made of a parchmentlike substance which is manufactured and repaired by a group of appendages designed for this specific task. The tube, about three-fourths of an inch in diameter, reaches a length of about sixteen inches from opening to opening; the worm is about nine inches. Usually both are smaller.

This tube is to the *Chaetopterus* what the shell is to the turtle; it cannot and does not leave it readily but spends its life constantly bent, quietly buried. Food and oxygen are acquired by the action of the cleverly designed suction pump which, working with rhythmic regularity, draws the water in at one end of the tube and expels it at the other. The micro-animals, small plankton, and other beings which are sucked in are filtered by a set of strainers located in the region of the animal's head and are then devoured.

It would seem that compensation in the form of safety would be a fair reward for the penalty of being imprisoned for life in a tube in the mud, but such does not seem to be the case. *Chaetopterus* is preyed upon by eels and perhaps by other fishes; and for the ultimate in ghoulish feasts the sight of a long writhing eel tearing one of these brightly lighted, utterly fantastic worms from its underground burrow must be the extreme. No one has ever witnessed this event but eels have been captured with their stomachs filled with the glowing heads of *Chaetopterus* and it is not difficult to imagine the scene that must take place.

In the dark of the night on the high tide the eels glide into the shallows, sliding serpent-like close to the bottom, nosing just over the mud. By some means, by scent, or the feel of the currents being drawn in or expelled from the tubes, or by touch, they discover the tube openings. Seizing the parchment in their mouths, they lash back and forth and tear the protective material apart. The exposed worms, highly excited, blaze with phosphorescence and retreat as far as their confined quarters permit. But not far enough, for the gaping mouths of the eels nuzzle deep into the mud, seize the brightly lighted heads; there is a brief struggle, a moment when the worms striving to retain their hold, are stretched taut in a desperate tug of war; then, unable to withstand the strain any longer, the head gives way and with a rending of tissues tears apart. The cold eyes of the eels and their glistening, slimy tissues must eerily reflect the light of the worms as they gulp down their prey. Small shining fragments, no doubt, float away in the tide, to be seized by the lesser fry waiting in the dark. It is not beyond the realm of reality to imagine that, for a brief time, while the phosphorescence lasts, the mouths of these preying eels gleam palely from the substance torn from the worms, and that for an equal period their stomachs glow as though they contain some sort of underwater electric bulb.

Most strange, so often do these worms lose their heads, and so accustomed are they to having them pulled off, they are provided with a special muscle to aid in the decapitation. When the head is seized and sharply pulled a ring muscle just in front of the sucker contracts and severs the extremity at the "neck." But this is no great tragedy, for the worms can temporarily do very well without it and the body quickly grows a new one. This is really a protection, for while the worm loses its head it thus saves its life.

To go undersea during a period of maximum phosphorescence is a never-to-be-forgotten experience, and there are few places where the midnight display is equal to that of the Chesapeake. The undersea at night in any conditions is exotic and foreign, but when the water is gleaming with luminescence and when great blobs of light suddenly flare before one's eyes like ill-defined jack-o'-lanterns and long streaks of fire burst past one's vision, only to fade in the distance, reminding one of the course of comets, it assumes a quality that is almost unreal. I have never felt quite comfortable during such times, although too fascinated to leave and return to the upper air and a more normal world. To stand erect on the bottom, or to hang suspended from a rope in midwater and to find one's body, and even the individual hairs on one's hand, etched in cold fire from head to foot is a queer sensation. One has the feeling of being a sort of living ghost, a misplaced submarine phantom.

As one becomes accustomed to the environment, as much as this is possible, and finds time for concentrated observation, the various lights take on individual character and a study of these will indicate their origin. The micro-lights, of course, dominate the scene. En mass they appear as a greenish or yellowish glow, and it is difficult to sort out the individual motes that make up the total. After some experience, however, the difference becomes apparent. The most common and the strongest micro-light is that of the *Noctiluca.* This lasts for some moments, perhaps for five seconds, before it expires; and it comes into being with a sudden flare that quickly reaches a climax and then diminishes. Occasionally when a *Noctiluca* becomes entangled it will light and remain lit for a full half minute or more, glowing steadily. In contrast, *Ceratium,* a strange little protozoan, with three long horns, is spark-like, and gives the effect of a star which bursts briefly into existence and then is extinguished. There are other sorts of micro-lights but these two are the only ones I am reasonably sure about; these I learned by examining water samples taken when one or the other type was dominant, as they sometimes are.

The most startling lights, however, are those of the larger jellyfish. The intensity of some of their luminescence is surprising. The comb jellies particularly contribute to the weird assemblage. One may be surrounded by their diaphanous forms, unaware of their existence, then suddenly, actuated by some stimulus, internal or external, or by contact, the dark is broken with brilliant bursts of livid green flame. The weird effect is heightened by the contrast with the dark and by the mistiness of the water, which partly obscures the more distant individuals and reveals them only as intangible halos. Often I have stood in the dark peering vainly for sight of some object when one of the comb jellies brushed against the glass of my helmet. It is difficult not to start when out of nowhere a greenish comb jelly comes into being only an inch or so from one's eyes.

The light of the comb jellies under usual circumstances, like that of a great many phosphorescent creatures, can be created only at night. Taken from the daylight into a darkened room they will fail to phosphoresce, or will glow so weakly as to be almost invisible. Nor will their luminescent powers return until they have been allowed to remain in the dark for some time or until their normal daily cycle has been completed.

Among the larger jellies which luminesce are the *Cyanea, Pelagia,* and *Aurelia,* the moon jelly. This last species is particularly beautiful and gives off an intense, bluish light. No one has ever been able to explain satisfactorily the value of luminescence to the jellies or to the other helpless plankton, nor how the development of their light organs began. It has been suggested that the light serves as a warning, and in those jellies which have long fragile tentacles, which might be torn apart by the blundering of fishes too large to be paralyzed by their poison, this power of luminescence might be of value. Certainly any fish which received the signal would sheer off if it had half a brain cell working. But this ready explanation does not help much in the case of the comb jellies, which have no extended tentacles. The truth is that we do not know much about the reasons or causes of light in certain animals, and it will require much long study and direct observation to ferret out the facts.

Once, many years ago, an interesting experiment was performed. An investigator placed a piece of fungi-infested luminescent wood in a glass jar and attached a suction pump. He withdrew a little of the air and was surprised to see the light begin to grow dim. More air was sucked out until the wood was in a near-vacuum. The glowing light disappeared. The pump was then disconnected and the air rushed in. Immediately the wood blazed forth again, as intense as ever. The experiment was repeated with a number of luminescent animals, with the same result.

This experiment established definitely that animal luminescence was a process of oxidation and that oxygen, either in the form of air or in solution in water or other liquids, was necessary. The glow created by a jellyfish, or a firefly, is as much a burning as the glowing coals in our fireplaces. The difference is that the light of a firefly, or luminescent bacteria, if extinguished, will light again once the needed oxygen is returned, even after a day or two, while that of the glowing coal once quenched is darkened forever. A second difference is in the amount of heat evolved. Less than one per cent of the energy required for bioluminescence is dissipated in heat. In an ordinary fire the heat energy loss is usually over ninety per cent.

The magic substance that is the fuel for these living, almost heatless fires, is a complex protein called luciferin and not phosphorus as was originally thought. Luciferin, so named after his satanic majesty, the keeper of fires, becomes lighted in the presence of oxygen and moisture when the combustion is promoted by an enzyme called luciferase. Luciferin itself is not sufficient, it must be aided by its ally chemical.

The organs that produce these substances may be simple or intricate. In some animals they amount to little more than tiny cells containing granules which produce the luciferin; in others, such as certain of the deep-sea fishes and decapods, the organs have layers of reflecting tissue, lens material to direct the rays and even devices to screen off the light at will like shutters on a window or, more accurately light-lids, resembling eyelids but having a reverse function. In most luminescent animals the light is controlled by nervous action and is turned on and off as the result of external or other stimuli. These stimuli are often received by whole groups of animals; fireflies have been observed to flash rhythmically and in unison, actuated by the same impulse; and such colonial organisms as *Renilla,* the Sea Pansy, have been known to show flashes over a whole group of animals at the same time.

In late June, there is a spot not far from the city of Annapolis, in Maryland, where I go if it is at all possible, each year to spend at least one evening. The place is a small isolated cove bordered partly by marsh and partly by a gleaming white beach. Back of the beach is a small meadow and in the back of this a dark pine woods. It is a scene that can be duplicated on the Bay in a thousand places. I found it one day quite by accident while skimming along the beach in a sailboat looking for shelter from a gathering squall.

The cove is a source of attraction because it possesses singular beauty, particularly at night, when its sheltered dark waters lie still and quiet and reflect the sparkle of the stars. The aged pines that guard it create a background for the pageant that occurs each year without fail, and their many needled branches etched in monotone against the luminous sky lend an artistic design for the annual show.

For each June the meadows between the water and the forest give forth legions of fireflies and for the space of several weeks, until the festival is over, the cove is a veritable fairyland. Literally thousands of the luminous beings flood the air and the sight of their comings and goings against the dark forest is truly an enthralling spectacle.

This festival of little lights is really a nuptial feast, the culmination of a year of preparation. During the days preceding the climax of the pageant, from down in the dank ground, from beneath the grass roots, there emerged hundreds of little beetles with brown and black wing covers. For a full twelve months these beetles had lived a grublike existence among the worms and moles; then within the space of a few days they were delivered from their old creeping forms, liberated from the soil; free to take soaring flight under the stars with lanterns to show the way.

But not for long. For down on the grass stems, on the petals of the flowers, and clinging to the leaves are the females. On and off their lights blink, glowing palely against the green stems. One by one the soaring, flitting beings in the air cease their restless darting, hover motionless for a time, and then glide down. On the tips of the grass stems the matings take place.

By mid-July the cove, and all the fields of the Bay

country, are devoid of glowing lights, the marriage festival is over. But down in the soil, down below the grass roots will be crawling thousands of tiny worms, burrowing, threading their way through the earth, the future fireflies, the gleaming meadow lights of the year to come.

CHAPTER 7

THE ANIMALS THAT GROW BACKWARD

LET US PICTURE A CHILD newly born, with all the potentialities of its race. And then, further, let us imagine that this child of our mental conception begins to grow and develop, that it shows evidence of intelligence and the promise of a fruitful future life. For a time it follows a normal course, assumes some of the characteristics of maturity, achieves mobility, is able to run about and indulge in the activity of its kind. Then suddenly, when it has seemingly progressed to a state of self-sufficiency a strange thing happens. The child becomes rooted to one spot, loses its ability to move, becomes flabby and rotund; its legs and arms shrivel and disappear; the eyes atrophy and lose their ability to see; the head merges with the body until it is indistinguishable; only the mouth, wide and distended, remains to recall the lively being that once existed. From a creature replete with senses and organs of locomotion it is transformed into a dull blob of immobile flesh capable of little more than an ability to take in food, digest it, and cast the residue away.

It is, of course, an incredible thought and yet one which is comparable to a similar transition in the life of a creature of the Bay.

Somewhere back in the dim ages of prehistory, long before the great dinosaurs and their kin roamed the earth, when the Appalachian Mountains were oozy bottoms of the warm archaic seas, there lived some insignificant-looking organisms. They were mere animate bits of flesh and tissue more or less like the countless other types of invertebrates that crept or crawled, swam or burrowed through the mud and ooze of the sea bottoms. But they differed in one respect. They had developed a notochord, a central nerve ganglion, the forerunner of the backbone, the spinal chord and the brain. We believe certain of these insignificant beings evolved into fishes, not fishes as we know them today but nevertheless fishes. And from these primordial fishes there developed through the ponderous march of the ages the amphibians, the reptiles, the birds, and the warm-blooded animals, culminating in man.

However, the determining factors that shaped these early creatures did not touch them all alike. Certain types went forward; the progress of their evolution is preserved in the rocks for all to see. Others persisted unchanged through the ages, generation succeeding generation in countless progression leaving them nearly as before. Their history is not so clear, for they possessed no bones, nor teeth, nor plates or scales to be compressed and delineated in the silted layered stones. But the Chesapeake Bay and all the waters of the world contain some millions of their descendants of an aberrant type. And if we may accept the evidences of ontogeny their evolutionary course is almost as plain as if there were a long series of fossils imprinted in the foliated strata.

We call these backward creatures tunicates, sea squirts, or ascidians, according to our preference. Every wharf piling has its colony, every rock or boulder carries at least one or two individuals and, along with the barnacles, they frequently share space on the bottoms of ships which have remained too long in the water and have become fouled. Although relatively few persons appear to be aware of their existence, the sea squirts are among the most common of the Bay's inhabitants. By their very nature they are unassuming, plain, and unobstrusive. But we may not look upon them with disdain, for their history is equally as long, if not as meteoric, as our own.

Strangely, each generation of these creatures repeats, in a sense, the story of their remarkable history, the early development of great promise, the forming of the notochord and the beginnings of a brain, the start toward higher things, and finally the passing by of the great opportunity and a return to the ancestral way of living.

It is the adult tunicates, or sea squirts, with which we are most familiar. In size they vary from the roundness of a small pea to the bulk of a lemon. Were we to descend beneath the surface and examine a single individual we would find it roughly globular, somewhat in the shape of a teakettle, only a teakettle of translucent green or perhaps amber-brown. And to heighten further the simile we would discover at one side a spout—and nearby another—as though the pot was arranged to be poured in two directions. But there the resemblance ends, for, looking more closely, we see tiny particles being sucked in one spout and discharged out the other.

If we were to cut a sea squirt in half we would find a marvelously simple but efficient organism. First would be revealed the outer skin and the two spouts—the shell of our living teakettle. At the base of the upper spout is a latticed basket filling nearly the entire animal. This basket serves a dual purpose and is an ingenious device. Few other living creatures possess anything shaped quite like it. For the basket acts as a net or filter, a type of organic colander that strains out all the edible organisms from the water. At the bottom of this net is a short tube ending in a small but—from an ascidian's viewpoint—satisfactory stomach. The excretory waste mingles with the strained liquid that has passed through the latticework and is expelled from the lower spout.

However, it is not the fact that the animal is equipped with a strainer that is remarkable, for many deep-sea creatures capture their food on this principle. It is the singular fact that the tunicates also use their strainers for the purpose of a gill, a sort of branchial chamber to supply oxygen for the animal's maintenance. The walls of the lacelike sac are filled with hundreds of minute blood-vessels; these, by the throbbing of a tiny capsulelike heart, carry the newly acquired oxygen on its limited path through the body.

Think for a moment of the significance of these seemingly unimportant statements. Although the tunicates are composed of little more than a shell, a heart, a stomach, and a set of reproductive organs, and although they are without means of locomotion or of action, sedentary, helpless blobs of green protoplasm, they breathe in the same way a fish breathes. And their similarity to fishes is further strengthened when a comparison is made with such forms as the herrings. For these fishes, like many others that live on minute swimming organisms, utilize a system similar to that of the foreign-looking tunicates. The herring takes in water through its mouth and from there passes it through its gill clefts. The gills strain out the floating, minute food particles and, at the same time, take up oxygen.

This point of resemblance assumes more importance when we consider that of all the strange and varied host of invertebrate animals that inhabit the land and sea, only the ascidians and their close allies possess such structures. Here is established a tangible link between the sea squirts and the fishes and all the other vertebrate animals up to and including man. It is not so farfetched as might be supposed to call a tunicate brother—or at least a cousin—even though the relationship is some hundreds of millions of years and an equal or greater number of generations removed.

However, it is not the possession of a combined net and lung that makes the tunicates most remarkable. We should consider the early stages of this simple creature. Sea squirts are anchored in one place for most of their lives; they cannot move about to find their mates and propagate their kind nor, like plants, can they cast seeds to be distributed in new places by the wind or tides. So they have solved the problem of ensuring the spread of their race by hatching free-swimming young. Other sedentary animals, the lowly oyster, for example, have adopted similar expedients.

The larvae, however, bear no resemblance to their prosaic dull-looking parents. Instead, they more closely resemble the larval tadpole stage of the frog; they are much smaller, the largest being less than an eighth of an inch in length. Like the tadpole, the immature tunicate possesses a heart, gills for breathing, a mouth, a stomach, and a sucker. Also, like the tadpole, it possesses a tail. Above all it has a primitive brain, a notochord which is the forerunner of a vertebral column, and a nervous system. Few tadpoles have much more.

A comparison of the structure of a larval tadpole and an equally infantile ascidian is startling. Except for a few minor variations they are alike. There is one great difference, however, which is not revealed by any sign of anatomy. They follow different roads. Although both beings pass through a metamorphosis, it is in opposite directions. The tadpole acquires legs, feet, a skeleton, true lungs, and an urge to ascend to the open air for a life on the borders of some quiet pool, or even, as in some genera, such as the *Hylas,* high among the branches of the trees. The ascidian does exactly the opposite. It proceeds only so far and then turns aside.

After a period of free-swimming life a strange urge comes to the immature tunicate. It develops a desire to attach itself to something. A rock, an oyster shell, almost anything will do so long as it has a firm surface for the sucker located near the mouth. Many tadpoles do the same, particularly those species found in swift streams. However, there is this difference: the tadpole merely attaches itself to rest for short periods, to save itself from being swept away by strong currents. Once the tunicate fastens to an object, it is in place for the remainder of its life.

A marvelous change then takes place. The waving tail is slowly absorbed, becomes in time a mere stump and eventually disappears. The body alters, becomes globular; the mouth opening enlarges, becomes a spout; the excurrent opening adjacent to the gills moves upward and becomes another spout. The stomach and the heart migrate down to the bottom of the creature, towards that part which was once the head. This heart is an odd device. Not satisfied with pumping steadily in one direction, like that of any ordinary animal, it possesses the unique ability of stopping every so often and reversing its direction. Thus, the vessels which at one moment serve as veins carrying blood to the heart, a second or two later become arteries conveying the fluid in the opposite direction.

And what of the promising nervous system, the primitive spinal column, and the brain? They degenerate and disappear, leaving only a spot of nervous tissue between the spouts hardly worthy of the name of nerve. The gill chamber becomes the efficient and simple food-catching basket—and our creature is complete.

It started life with great promise. Time and the potentialities of its race left it as a simple breeding, feeding, stationary animal. Yet, paradoxically, it fills its small role to perfection. Within the confines of its tightly bound translucent body it reproduces its kind, secures its food, and with a minimum of effort, placidly lives out its existence, unhampered and unconcerned by the worries and vicissitudes of the hectic world it might have inherited if its peculiar destiny had not decreed otherwise.

Living side by side with the tunicates is another creature, one so commonplace that we give it little thought. That is, all except those few of us who have to do with ships and other seagoing equipment. To us the animal is an unmitigated nuisance; a devilish organism that costs many thousands of dollars a year and extracts hard-won earnings from thrifty maritime pocketbooks. I refer to that paradox of animals, the barnacle.

Contrary to popular belief, the barnacle is not a shellfish. Instead, it traces its family tree with the crustacea, that diversified class which includes the shrimps, the lobsters and the crabs. While it is true that the barnacle has a shell like a mollusk, nevertheless it is more closely allied to the bugs and beetles, which ravage our gardens, and even to the butterflies, which soar over the fields, than to the oysters, clams, or mussels, which it superficially seems to resemble. But then many of us pass for one thing when we are quite another.

Ignorance of this deception was not confined to laymen, for so cleverly had the barnacles concealed their identity that even the professional biologists for many years were unaware of their true nature. This is not surprising because the anatomy of an adult barnacle when dissected from its enclosing shell is most uncrustacean. Indeed, the animal appears to be little more than a blob of gray flesh with a reproductive system, a stomach, and some breathing organs, a mongrel among organisms.

One day, however, some inquisitive naturalist decided that it would be intriguing to study the reproductive system of these ill-defined animals. To his amazement the microscopic beings that hatched from the barnacle eggs were about as unlike the adult as could be. The newly born creature resembled some outlandish insect; it bristled with hairs, barbs, spikes, and antennae; queerly, it had no shell or any hint of one. The contrast between parent and progeny was so marked that if an armadillo had given birth to a butterfly, or an elephant to an aardvark, the occurrence would be no more strange. The creature swam and it was transparent, an odd thing of numerous segments, with a feathery tail and two large dark eyes all out of proportion to the size of the head. It was exceedingly small and a strong lens was needed to delineate all its details.

The inquisitive investigators named the glossy being nauplius and continued to watch. The nauplius in time molted, and then molted again, shedding its skin and each time changing its shape a trifle. By now it was obvious that the altering beast was a crustacean. Its body was beginning to acquire many characteristics of the young of other crustaceans and, in fact, it resembled more or less the young of certain stages of those more prosaic crustaceans, the crabs, the shrimp, and of the ubiquitous copepods and isopods. Then, oddly enough, and unexpectedly, instead of continuing its development in a sane, normal crustacean manner, the nauplius suddenly acquired two small oval shells, somewhat in the shape of a miniature clam, one for each side of its torso. The biologists, having the urge of Adam, felt that this new stage should be named. So they called it cypris after the goddess Aphrodite, who was alleged to have risen from the foam of the sea on the Island of Cyprus in the Mediterranean.

But more changes were still to come. The watching scientists were intrigued one day to see the cypris settle on its back and thrust its six, jointed legs into the air. Then another series of molts occurred and after each molt the barnacle-to-be was slightly altered. First, the two calcium shells were discarded and then were replaced by the conical shell of the adult barnacle with which we are so familiar.

But the most astounding change of all was in the legs. Instead of being absorbed as were most of the other organs, becoming fused with the bulk of the body or disappearing, the legs spread out, fringed apart and completed their metamorphosis looking like so many curled feathers. And so the barnacles are termed, scientifically, cirripeds, the feathery footed or hairy footed. These feathery feet interlace to form a sort of net. Thus a barnacle feeds. The feathery feet are thrust straight out of the shell, are cupped in net shape and then withdrawn, bringing with them the microscopic gleanings that go to make up barnacle food. The excess water escapes between the interlacings of the feather feet. These feet also bring in the necessary oxygen for the maintenance of life. As we human beings depend upon our efficiently shaped hands to provide our nourishment, build our homes, and fulfill our many tasks, so the barnacle depends on its feet.

Here we have a remarkable similarity in development to the sea squirts. Both start life as free individuals able to swim and move at will; both complete their life cycles as stationary, inert organisms living under identical environments, the same end attained by divergent paths.

It is a paradox that the barnacle, so sedentary and inert, is yet a world traveler. Although, once it has reached the adult state it cannot move an inch from its place of fixation, it visits the oceans of the world on the bottoms of tramp steamers, on the backs of whales, even on the umbrellas of giant jellyfish. One strange form is even found on the fins of flying fish, so it may be said that these intrepid voyagers were accustomed to air travel and made it a habit some thousands of years before the Wright brothers performed their curious experiments at Kitty Hawk.

The barnacles are protected by an ingenious device. Located near the apex of their conical shells are two pieces of lime known as the terga and the scuta. These are movable and may be compared to sliding doors. Should some unfavorable circumstance confront the animal, it merely moves the terga and scuta together and excludes the evil from its dwelling. The two pieces mesh perfectly, sealing the creature in its shell until, once again, it desires to open. Thus the barnacles can survive low tides when they are helplessly removed from their native element. The doors are quite watertight and they retain all the moisture needed to prevent dehydration until the tide returns to normal.

Our Chesapeake barnacles of today are small affairs; few of them exceed an inch from base to apex. But there was a time some forty million years ago when the waters that rolled over the Bay region were filled with giants of barnacles, veritable Brobdingnagians among cirripeds. Their remains are well preserved in the fossile strata of the Miocene deposits in the cliffs along the present Bay front of Calvert County on the Western Shore of Maryland. These old-time crustaceans were no less than four to five inches long and up to two inches wide. Not long ago I picked up a group of these mammoths, joined together on the shell of a mollusk which has long been extinct. The shell had fallen from the cliff into the water and had lain there for a long time before I found it. Clustered over both the mollusk shell and the cases of ancient cirripeds were a dozen brand-new living barnacles, almost identical, except for size, with the giants on which they were resting. The contrast in dimensions was striking; but, nevertheless, the important fact remained that, although the mollusk had been overwhelmed by time and the circumstances of geologic change, the barnacles by their very versatility and the ingenuity by which they had forged a unique niche for themselves in their world of the deep-sea have weathered the vicissitudes of uncounted centuries and the passing of several geologic ages. It is questionable if man, with all his flair for invention, his artificial aids to living, and his present domination of the continents will be thriving some forty million years hence. For their pertinacity, if nothing else, barnacles deserve our admiration, perhaps our respect.

CHAPTER 8

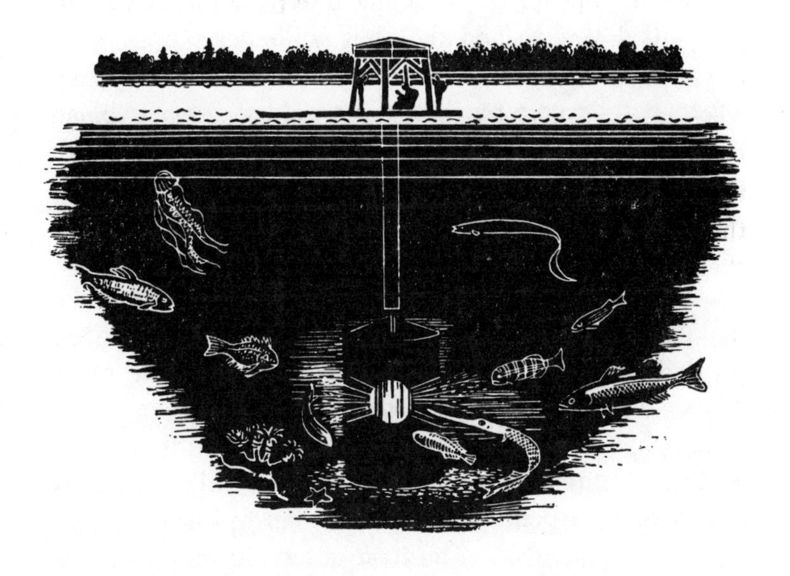

THE BENTHARIUM EXPERIMENT
Twenty-four Hours Beneath the Bay

THIS EXPERIMENT HAD ITS INception one day when I was diving in a helmet near Point Lookout, Maryland. I had been down on the bottom for nearly half an hour watching the actions of some blennies which were behaving in a strange manner. These blennies are queer little fish and they were maneuvering over the bottom in pairs as though engaged in a sort of submarine dance. In addition, there were a number of other fish; it was one of those rare days when the Bay floor was astir with activity. For a long time I lay still, trying to glean some hint of the meaning of the performance, but presently I could stand the chill no longer.

The cold had penetrated every fiber of my body; my fingers had become pinched and wrinkled, washerwoman style, and they looked white, even a little blue. With teeth chattering I had to go to the surface. On the way up I blundered into a large jellyfish, which wrapped itself around my shoulders and left a burning memento to contrast with the cold. Frustrated, I lay over an hour in the sunlight before I felt warm enough to go below again. By that time the blennies were gone and all the other fishes with them. The sea floor was deserted.

The physical discomforts of helmet diving in the Chesapeake are considerable and the regulation diving suit is cumbersome, expensive, and not the most pleasant device in which to spend an afternoon observing undersea life. In either case there is the old bugaboo of pressure. In shallow water this is not important but in more than forty feet the head tends to reel; after half an hour the slightest exertion becomes laborious; a heavy hand seems to press on the chest and stomach. Eventually there is no choice but to give up and go to the surface.

After the Cove Point experience I began casting about for some simpler and more comfortable means of going undersea and at the same time remaining dry and warm, breathing fresh air at normal atmospheric pressure, and staying for indefinite periods. The obvious answer, of course, was some glass-windowed sphere or cylinder large enough to hold one or more persons and which could be tightly sealed against the water. Such an idea was not new nor particularly startling. Mr. Beebe had developed his famous Bathysphere and Mr. Williamson his submarine tube. Both of these devices, however, while very efficient, were exceedingly expensive and required considerable equipment to operate them. At that particular time the depression was on in full swing and funds for such ventures were difficult to secure. Therefore, whatever was to be devised had to be inexpensive.

I confided my scheme to Dr. R. V. Truitt, director of the Chesapeake Biological Laboratory at Solomons, Maryland, and to a group of associates at the Natural History Society of Maryland. Dr. Truitt offered the facilities of his, then new, laboratory and the men of the society contributed materials, suggestions, and labor. A tentative design was drawn for a cylinder of steel about four feet in diameter and six feet in height. A large window was needed for observation and photography. It was estimated that the cylinder would accommodate two men comfortably, three if necessary.

Securing the equipment was the greatest headache. The cost of a steel cylinder manufactured to design was found to be prohibitive. Fortunately, someone dug up an abandoned piece of industrial equipment which, with some alterations, almost exactly fitted the prescribed dimensions. It had been used as a container for some sort of chemical, exactly what we never discovered. On the side and at precisely the right location was an opening twenty-two inches in diameter which would make a splendid window.

A manhole was cut in the top and a cover provided with iron dogs to fasten it tight. Fitting the window turned out to be quite a problem. At first we decided on a pressure seal with rubber gaskets. The glass, a circular piece three-fourths of an inch thick, was laid in place and the bolts were carefully screwed up. One after the other they were given a quarter turn to bring the gasket down evenly. The next to the last was being tightened when there was a loud cannonlike report. To our dismay the glass was split exactly in two. A second piece fared similarly in spite of all precautions. Eventually we solved the problem by redesigning the window and floating the glass loosely in a circular channel filled with white lead. This worked successfully and the window never leaked.

Two air lines were fashioned, one for incoming air, the other for exhaust and attached to an air pump constructed from an old motorcycle engine. This proved to be quite efficient and later we found that we could even smoke cigarettes without fouling the air. The cylinder was under ordinary atmospheric pressure and at no time did we feel any discomfort. For safety the ends were equipped with valves which could be shut quickly in case one of the lines parted.

We were considering telephones but found them unnecessary because the exhaust line functioned as an excellent speaking tube. For convenience we attached a short length to the valve inside the tank so that we could sit comfortably and describe events to those on the surface.

To support the equipment, and to raise and lower it, we built a rectangular float fifteen by eighteen feet with a well in the center. Over the well was a quadrupod derrick supporting a chain fall which in turn was attached to a steel ring welded to the cylinder.

The problem of ballast was solved simply and inexpensively. The weight required to overcome the natural buoyancy of the cylinder was about three tons. Fine, close-packed sand stowed in small muslin bags provided all the ballast we needed. A large number were stowed in the lower half of the cylinder where they served the double purpose of ballast and seat. The balance were hung on a steel ring welded to the outside.

By calculation, it was estimated that the equipment would withstand pressures down to a depth of three hundred feet with a comfortable margin of safety. Since few places in the Bay went much beyond one hundred fifty feet, the entire region was open for exploration. It was no longer necessary to be concerned about stinging medusae or penetrating cold. At the greatest depths we could sit comfortably and as long as we desired. When we finished, the total cost of the tank, derrick, float, and accessories, not including the value of donated or loaned equipment, was less than four hundred dollars, a very modest sum.

We felt that the device should have some sort of name and after numerous suggestions decided to call it the "Bentharium" after the Greek terms *benthos,* referring to the bottom of the sea, and *arium,* a room or place. For lack of champagne, the christening was done with a bottle of Coca-Cola smashed over the edge of the window.

The Bentharium was built in a little creek at Solomons Island, Maryland, and when the motorboat arrived early one sunny morning to tow us into the broad reaches of the Patuxent River for the first trial, I am sure that the then-peaceful bayside village had never seen such a queer contraption pass through its pleasant little harbor. The Solomons Islanders, for the most part, were noncommittal, but a goodly number were quite certain that before long there would be a first-class drowning.

And, to be honest, as the entrance of the harbor was passed and we went out into deeper water I was not so sure myself. The thought kept recurring that the big glass window would break under the increasing pressure. At one hundred feet the weight would be sixty-two hundred pounds per square foot. In my mind I could picture the sudden surge of green water and the horror of being trapped. As an afterthought I threw a small sledge into the opening to beat out the jagged fragments if that became necessary. On calmer reflection, however, I reasoned that even if the glass did give way, enough air would be retained in the upper portion of the chamber to keep one alive for some time.

We anchored in fifty feet of water about a quarter of a mile from shore and prepared to go under. The Benthario's crew consisted of five volunteers, I. Hampe, J. White, L. Putens, F. Yingling, all of Baltimore City, and myself. Straws were drawn and Putens and I won the deal. White took charge of the air pump, Yingling and Hampe handled the chains and remaining ballast bags. We "lucky" ones climbed in and seated ourselves on the sandbags. A second later the light streaming in through the opening was cut off and with a dull clang the lid was dropped in place.

For a brief second there was a feeling of being trapped, as indeed we were, for entry from the outside was now impossible. The bolts that held the hatch cover were all fastened from the inside. We spun the nuts tight, made certain they were secure, and then turned to the window. It was half in and half out of water. Through the upper portion we could see the woodwork of the float, the tops of the waves, patches of blue sky and white clouds, and in the distance the red and white buildings of the laboratory. The lower half showed pale green tinted with yellow, shading to darker colors below. Of most interest was the dividing line, that fraction of an inch which separates the world of the upper air from the cool depths. It quivered and danced; the underside was pure silver and resembled molten mercury. A surprising amount of life was clinging to this opaque ceiling. Tiny larval fishes, hardly a quarter of an inch in length and consisting mostly of big eyes and transparent tails, swam just underneath. They yielded to every quivering motion of the ripples; their lives were one continual agitation, an incessant swinging back and forth. Yet in their diminutive manner they moved independently, scurrying in small circles, seeking other small beings for their nourishment. The quantity of these baby fish was astounding.

Mingled with the fishes were other motes, minute crustaceans, numerous small round objects which appeared to be egg cases, eighth-inch jellies and a host of smaller organisms which, because of their microscopic proportions, defied identification.

A voice came down the speaking tube asking if we were ready. We replied that we were and to go ahead and lower away. A moment later there came a hollow grinding of chains and with a queer lurching movement we began our descent into the depths. The water swiftly mounted the glass and soon engulfed it entirely. Presently we heard the waves sloshing over our heads as the hatch cover came awash, then this suddenly ceased. Peculiarly, however, all the other sounds were as plain as ever. We could hear the banging of things on the float, footsteps on the deck, and the gears of the chain fall.

At six feet we called a halt and inspected the equipment. The window seemed tight and so did the hatch. Except for a condensation of moisture along the steel walls we were quite dry. Satisfied, we looked out the window again. The surface was still visible but it had acquired a misty look, a sort of lustrous sheen. Try as we might we could not see beyond it; it shut off the world of the upper air as completely as if it had been a metal curtain. Only the understructure of the float gave a hint of what was going on above. In the distance the anchor rope pierced the ceiling and curved downward out of sight.

The most interesting phenomenon at this point, however, was not the surface but the rays of light from the sun. These were striking downward in long shafts, brilliantly high-lighted for a few seconds; then they faded only to be replaced by others. None remained fixed for long; there was a continual progression, some wide, some narrow. Each shaft during its brief existence acquired a pearly misty quality, a soft iridescence which was accentuated by the dark color behind.

For some time we watched the display and then asked to be lowered slowly and to be kept advised of the depth. At ten feet the shafts were barely visible, at twelve they were gone. The display was restricted to the upper zone. The change in color at various depths was marked. For the first three feet the water was golden, suffused with sunlight; then the gold disappeared except for the dancing rays; from six to ten feet these faded rapidly and green became predominant. From ten feet to the depths the green became more and more solid, more somber, darker. At first it was lively, bright and pleasant to view but as the depth increased it altered to deep olive and turned cold and black.

The voice came down the speaking tube again—twenty-five feet. With faces pressed against the glass we looked about. The water was quite blank, there was not the slightest suggestion of life, nor was there any visible object to fix direction or focus. Then came the only real scare we had during our Bentharium diving. From somewhere we heard a trickling sound, as of water suddenly splashing, a hollow gurgling noise. It sounded as though a liquid flood was rushing in upon us. Hurriedly I grabbed the speaking tube and asked if everything was all right on deck. We were told that it was—but that a ferryboat was passing close by on its way across the Patuxent.

So that was it. The propeller was causing the gurgling and splashing sound. Relieved, we relaxed and heard the noise reach a climax, then slowly fade away. Later we were able to identify a number of boats by their sounds. Ships a half mile away could be detected easily through the sounding-board sides of the Bentharium.

At thirty feet the first fish came into view. They were only little fellows, graceful objects with a broad band of silver running down their bodies. Appropriately they are called silversides. They were accompanied by some killifish. We were a little surprised to see these because we had always thought of them as creatures of the shallows. Commonly they spend their time within a yard or two of the shore, where they live in myriads.

The killifish and the silversides remained with us all the rest of the day. For some reason the Bentharium seemed to fascinate them, particularly the window. They spent long minutes vainly trying to swim through and they never seemed to get it into their heads that it was impenetrable. Frequently they rushed headlong into the glass only to stop with a sudden bump. We could almost imagine the look of pained surprise that should have come over their immobile features.

We stopped for a while at thirty feet to watch the small fry. While we were thus engaged a big *Dactylometra* drifted by. For once we could see it pass with equanimity. The medusa drifted through the swarm of silversides and they parted to give it passage. When the jelly went by the dangling anchor rope, one of the tentacles dragged over the strands. We were interested to observe that it was quickly contracted and drawn out of harm's way.

The fish followed us as we went down again. Forty feet passed, then forty-five. A second later we heard a grinding, crushing noise beneath; outside a cloud of brown silt rose and slowly floated away. We were on the bottom. To our surprise, although the lower edge of the window was only two feet from the Bay floor, we could not see the sand. Some trick of refraction, caused by the angle of the glass or of the water, prevented vision downward. Later we discovered that the only way to view the sea floor was to tilt the Bentharium at an angle.

We sat on the silt for perhaps half an hour. As soon as the grinding of the chains ceased, everything became very quiet. No boats passed to break the silence, nor could we hear the waves slapping on the float. The tide that was sweeping by on the surface at fifty feet down was barely moving. The only motion was that of the minnows which, contrary to all expectations, were still clustered about the window.

Before descending we had prepared a bag of crushed crabs to attract whatever might be in the vicinity and had acquired also the carcass of a long-dead croaker to lure crabs. We asked that these be lowered in front of the window. They suddenly appeared surrounded by a host of excited fishes of the same types that had followed us down. The crab meat seemed to drive them frantic and they snatched and tore at the flesh as though they had never had a good meal. In a few minutes the bait was gone and they were beginning to mill around the less desirable croaker. But suddenly they scattered and the cause of their flight was soon revealed, for a large male blue crab came sidling out of the haze and attached itself to the dead fish.

We watched it perform a clever surgical operation. With its big claws it slit open a line across the fish's abdomen, then another at right angle to the first, and finally a third parallel to the first. The jaws cut the tissue like a pair of scissors, effortlessly and without jagged edges. This accomplished, the animal peeled back the square of flesh as one would open the cover of a book. It was all so neatly done, we were surprised that a creature as "brainless" as a crab would proceed so sensibly.

The crab began feeding on the uncovered abdomen and after watching the saprophagous feast for a time we signaled to go to the surface. We arrived just in time to receive a bevy of newspaper reporters who had been assigned to cover the descent. So we spent the next several hours giving each a trip down into the Bay. Fortunately, the silversides and a crab stayed around, so we could show something. Later, we learned that hours could pass without a living object going by our window except the ever-present jellies. Even with the most delectable bait, the Bay at times appeared a watery desert, devoid of all activity.

We had such a session after our visitors left. About sunset a strong wind came up and before long was blowing a gale. After supper we rowed out into the river, lit a riding light, and lashed the equipment down. The Bentharium was banging and thrashing about at a great rate, so we lowered it partway into the water to ease the strain.

During the night the gale developed into a howling southeaster and we spent a sleepless night keeping watch and worrying. In the morning the wind was as strong as ever. Among the things that interested us were the actions and habits of various organisms during stormy weather. Many fishes, for example, cease feeding during certain heavy winds; we were curious, also, about how far down the wave disturbances went and what the jellyfish were doing. When the sea gets rough they commonly desert the surface, which they seem to prefer, and go down to less agitated regions.

So we determined to make a descent in spite of—or rather because of—the weather. When we reached the float we found waves sweeping over the decks and the Bentharium bobbing up and down like a cork. While we loosened the fastening a steady rain began to fall. A drearier or more gray day would be hard to picture. Joe White and I dropped inside and soon had the lid on and bolted although we nearly got seasick doing it. The air inside was stuffy and there was a heavy reek of oil; the motion was nauseating. For a few minutes we had a lively time. The deck crew had their hands full trying to keep the window from smashing into the float.

"For Pete's sake lower away before I get seasick." I called up the speaking tube. I swallowed hard to keep my breakfast—and won after a struggle. To lose it in such confined quarters would be too horrible to contemplate. I heard Hampe at the pump laugh and make some smart remark, but shortly after the chains began to rattle in the fall.

We turned to the window. The water had lost all the lovely yellow color of the day before and showed a dead green. Big waves smashed against the glass and dissipated in great silvery masses of bubbles which sprayed downward, danced crazily about, and then raced for the surface.

Before long we were below the bubble area and, to my relief, the Bentharium became more stationary although we could still hear the waves breaking against the float. The window then showed blank green. There was a large amount of material in suspension, hundreds of particles of solid matter to the square inch. At ten feet we could still see slight evidence of the wave action by the rhythmic shifting of suspended particles. Below this it was not apparent due to the fact that we were bobbing up and down ourselves with the movement of the float.

At about fifteen feet down we noticed a tide running by that was not apparent on the surface. Until this time we had seen no life. Then *Mnemiopsis*, comb jellies, began to drift by in large numbers. We called a halt for a while, then went down again. At twenty-five feet they thinned out. We had ourselves pulled up again and found that the jellies were segregated in a stratum about ten feet thick. There were comb jellies above and below this stratum but as scattered individuals only. Within the band there were thousands.

Except for the comb jellies, although we stayed underwater several hours during the storm, we saw no other life. The fishes, crabs, prawns, and all the other creatures were hidden or had secreted themselves on the bottom or in little nooks and crannies away from our sight. Although we had all kinds of bait it did not bring them forth. We can offer no explanation and it will take many more descents into the Bay before we can guess at this or many of the other questions that presented themselves.

One of the objects of the first season's work with the

Bentharium was to remain below the surface as long as possible during the entire course of a day. We were not interested in doing this to establish any sort of record or to see if it could be done; instead, we were concerned with the fluctuations of life and the changes of scene that took place as the daylight slipped into night and night back into day. We were also curious about the effect of the tides on the life of some of the Bay's creatures.

When the storm abated we arranged a series of shifts, each lasting about three hours. We started our experimental day promptly at ten o'clock on the morning following the storm. Fortified by a large breakfast and laden with a goodly supply of sandwiches and iced soft drinks we went below for the first vigil.

The water had cleared surprisingly fast, considering how it had been stirred by the gale. By noon it had regained its pearly, misty, liquid green color which we had found so lovely on the first day. The comb jellies, no longer restricted to a narrow band, had scattered again although most of them had migrated to the surface, which was now quite calm and still.

As we had found in our helmet diving, events occurred without warning. The fishes we saw came in bunches, appeared suddenly, remained for long or short periods, and then vanished as quickly as they came. Many more creatures stayed just out of vision. The most baffling circumstance about diving in our Temperate Zone waters is the mistiness. Just beyond the limit of clear vision, only six feet or so away, one discerns vague shapes or sees the glint of scales reflected from finny bodies, senses the presence of hundreds of schooling fishes, or sees the shadows cast by their bodies. But they are just out of reach; their blurred forms pass, move or weave about, but no identification is possible. At times it is maddening to see form after form hurtle by or drift slowly past on the tide and not know what they are. Although the creatures within close range are clearly visible and even appear slightly magnified, those beyond six feet are half obscured.

During the daylight hours we were visited by many organisms but nearly all of them remained just out of reach. Some few were recognizable. Once a great school of menhaden poured by. We were able to identify these by their peculiar golden color and their odd mode of swimming. When in a hurry, as these were, they appear to vibrate rather than undulate and when the light catches their scales and reflects it it acquires an unusual pulsating shimmer which is unmistakable.

Someday, when time permits, I am going to prepare a key for quick identification of fishes in their natural surroundings. We have found that when one goes underwater one must abandon all preconceptions of the colors of fishes. Fishes which in the upper air are plain gray or silvery, when viewed in their own medium may become the colors of pearls, brilliant yellow, old rose or iridescent green or blue. Thus menhaden, dull things of tarnished silver, become glittering objects of burnished coppery gold. Anchovies, those little fellows which are the source of the dismal mess known as anchovy paste, are really flaming lavender. Mullet, ordinarily dull mottled gray, assume a deep purple edged with yellow when lighted at oblique angles by the sunlight.

I have talked with many fishermen, oystermen, and other watermen, and of all these none seem aware of this considerable difference in color. In some cases the alteration is so great that the only comparison I can think of would be to assume that all the birds of the upper air had suddenly changed hue and that sparrows had inexplicably been transformed from dull brown to iridescent grass green; or that bluebirds had overnight become saffron yellow or mockingbirds deep scarlet with purple primaries. That this is not so exaggerated as it sounds will be attested to by anyone who has spent any time undersea.

The change in color is most marked at night when artificial light is the only source of illumination. Then the display can become utterly gorgeous. We were treated to such a scene shortly after the passing of daylight.

The day slipped by almost imperceptibly. The green water became gray, then blue-black, and finally completely dark. For a long time we sat still watching the minute "exploding" of phosphorescent *Noctiluca* and other protozoans. Frequently long trails of light marked the passage of a fish or the hurrying of a school.

We were equipped with a battery of lights consisting of a series of four 500-watt bulbs mounted in a watertight case. Laboriously we had laid a lead-covered cable out from the laboratory and, after ruining a couple of fuses and nearly electrocuting ourselves on the wet deck, succeeded in getting it working. When it was turned on it burned with a fierce white glare. Then, whenever a streak of phosphorescence showed in the dark, we threw the switch and attempted to identify our visitor. Usually all we got for our pains was a flare of color—pink, blue, green or crimson, or a combination of all four—before the creature was out of sight.

Contrary to everything we had expected, most of the larger fish avoided the light as though they were afraid of it. For the smaller beings, however, it was sheer fascination. The anchovies, particularly, could not resist it.

They clustered about the bulbs by the hundreds and although they individually were only an inch or so in length the massed color of their iridescent pink-lavender sides filled the vicinity with a cerise glow.

The light also attracted a number of halfbeaks, those singular elongated fishes with a curiously extended blade on the lower lip similar to the sword of a marlin. The tip of this member, unlike the weapon of the swordfish, is quite soft. It is brilliant scarlet and this, contrasted with the beautiful glaucous greens of the remainder of the body, is quite striking. On the surface these fishes are plain silver. Although it is assumed that the soft tip of the lower mandible is a sensory organ, these particular fishes gave the bill a terrific pounding against the glass of the light box in their frantic efforts to reach the light. One individual rammed so hard that its bill broke halfway down, and hung at an odd angle. This unfortunate, however, did not seem too inconvenienced and stayed around as long as the light was lit.

Later in the evening we made an interesting discovery. Several of our bulbs burned out and it seemed that as the glare diminished the number of visiting organisms increased. So we extinguished the light and substituted an ordinary flashlight with a focusing beam. The number of visitors doubled. The nocturnal creatures were much more attracted by a low-power solitary light than by an intense glare. After this we abandoned the light box and its troublesome cable.

Toward two in the morning we were deluged with worms. At this late hour most of the fishes had gone elsewhere; even the myriad prawns were missing. The worms appeared to be small *Nereids*, bristly affairs with a multitude of segments and appendages. They were out of reach and we were unable to secure any for identification. There had been a few about all evening, but from somewhere they suddenly and mysteriously appeared by the hundreds. They were reddish in hue, slightly iridescent, and were endowed with incredible vitality. Their bodies vibrated as if by unseen electric currents, as if they were all seized with an exaggerated St. Vitus's dance. In sweeping circles they went scooting through the water, quivering in annelidian frenzy, turning over and over in their courses, executing the most intricate figure eights, loop-the-loops and Immelmann turns. Occasionally a group would get together and, bunched in a tight sphere, go whirling round and round. At these times their energy doubled until their bodies moved so rapidly that they appeared as only crimson blurs.

These marine worms, and others of their type, have always been a source of amazement. They are so unlike the good old-fashioned conception of a worm and their ways are so inexplicable that I cannot make sense of their antics. I know of no living creatures, except possibly the hummingbirds and certain insects, which put so much action into their movements. Whenever we have seen them they have been bursting with energy. It is possible they were breeding and that their excited swimming was a part of their nuptial ceremony. Normally—at least it is so supposed—they spend the greater part of their lives buried in the ooze and silt of the sea bottom. Yet, there was rarely an evening when we did not see their grotesque forms in numbers in the open water. From direct observation I am inclined to believe their time is divided: that during the day they sleep in the black crevices of the Bay floor and that they emerge during the night to undertake whatever queer tasks and habits fall to their lot. On this particular night their frenzy lasted until morning, when they mysteriously disappeared.

The climax of the day beneath the waves came at dawn and came so unexpectedly that we were not prepared for it. By four o'clock we were all tired and even the creepy spectacle of the gyrating worms had lost much of its interest. The deck crew were lying around half asleep and down in the Bentharium we could scarcely keep our eyes open. Our shift was nearing its end and in anticipation we had ourselves drawn up to within twenty feet of the surface. Just before daybreak we turned off the lights. At first our eyes failed to register. Then through the window came the familiar little pinpoints of light flashing on and off like stars. Some comb jellies went by with their bands of cilia brightly glowing. The intensity of their light was surprising.

Presently the lights began to pale, ever so gradually, until only the brighter comb jellies made an impression on us. Then we noticed for the first time that the rim of our window was dimly visible. Soon the faintest gray sheen came visible through the glass. But what a gray! Tinged with the most evanescent of blues, it was unearthly in its shimmering, opalescent quality. In all our descents we had seen nothing like it. As we watched, the blue became stronger and then slowly, almost imperceptibly, altered to an indefinable gray-blue-green.

We asked the deck crew how light it was and received the answer that the shore line was plainly visible but that there was no rosy tint in the east as yet. The inside of the Bentharium was becoming visible and we could distinguish our hands and other objects. Outside all traces of phosphorescent organisms had disappeared. Once again we turned to the window. The blue tone of the light had entirely vanished and the water was assuming a pearly green color somewhat like and yet unlike the green of full daylight. We supposed that the change of color was over. The greatest surprise was still in store for us.

At that moment the deck informed us that the edge of the sun was coming over the horizon. Down below there was no inkling of it but in five minutes the edge of the window began to assume a pinkish tone and shortly after a deep orange-red. Then rapidly the water itself began to alter. Yellow-green at first, then deep yellow, orange, and finally orange-red. Looking upward through the window we then saw the most marvelous display of all. Hurtling down through the orange water came great red balls of light, murky with diffusion and awe-inspiring. They were the red images of the sun rising above the horizon in a great crimson ball focused and shot downward by the facets of the waves.

I looked at Hampe crouched beside me. He was bathed in orange light which flared and dimmed as the balls reached our windows. We were nearly speechless with the unexpected beauty of it. We had been down at sunset but had witnessed no comparable scene. Then swiftly the orange turned again to yellow, yellow-green, and finally the brilliant translucent green of daylight with which we had grown so familiar. We gave the signal to come to the surface.

Throwing up the lid of the hatch we were greeted by the sun well above the horizon. Over near the island some gulls were diving after menhaden and screeching their usual morning chorus.

CHAPTER 9

POINT OF THE POTOMAC

IN NUMEROUS CASES THE SELF-appointed Adams who were responsible for the designations of the various points, creeks, inlets, and rivers of the Chesapeake showed a fine sense of nomenclature. What could be more intriguing, for example, than Point-No-Point, the name of a blunt promontory near the mouth of the Potomac, or more descriptive than Bloody Point at the southernmost end of Kent Island, the scene of a piratical murder in colonial times. Many of these more appropriate names are worthy of description and have been treated in detail in other, more historical works about the Chesapeake. Thimble Shoals, the Hole in the Wall, Wolf Trap, the Rip-Raps, Dames Quarters, are only a few of the more engaging.

However, in most instances the namers of the places on the Bay failed to describe adequately the localities they placed on their maps; they exhibited an astonishing lack of imagination. The Bay is full of names as meaningless as they are prosaic. James Island at the mouth of the Little Choptank on the Eastern Shore is, or rather was, one of the most entrancing spots on the entire Chesapeake but its designation gives no hint of the wonderful old pine trees that, straight as the masts of ships, line its shores or of that exquisite little inlet which, alternately bordered with white sand and edged with green marsh grass, cleaves it almost exactly in twain. Neither does Thomas Point at the mouth of the South River suggest anything of the character of the peninsula bearing that unromantic name, of its magnificent oaks and gum trees, or of the eagles that have nested there for years, or of the quiet duck-frequented cove that it shelters from all but the heaviest gales.

The colonial geographers, for the most part, were doughty and estimable characters but there was no sense of the artistic ingrained in their adventurous souls. Their classification of the Bay's places is a succession of good old-fashioned English names, dull as they are misdescriptive. Only the nomenclature inherited from the Indians bears a full flavor and a sense of rightness, although, for the most part, the meanings have been lost.

Thus it is not surprising that one of the most interesting areas in the entire Chesapeake should have, for these long years, escaped notice. Only a handful of naturalists are aware of the character of the locality and few have committed a description to print. It is small wonder, for it bears the incredibly plain name of Smith Point.

Undoubtedly it was named after that amazing and garrulous character Captain John Smith, colonial explorer, mapmaker, romancer, and storyteller par excellence; this fact alone excuses the misappellation.

Viewed from the Bay, Smith Point does not appear different from any one of several hundred promontories which front on the Chesapeake, nor is there any detail of its outline which might cause one to stop and investigate. Its secret, like the worn cover of an old but good book, is hidden beneath a deceptive exterior.

Although I had some previous knowledge of Smith Point by hearsay, the circumstance of my first visit was dictated as much by considerations of finding shelter from a high wind and from an exceedingly rough sea as by an interest in natural phenomena. All day a half gale had been blowing from the northwest and although it was mid-June its intensity approached that of later or earlier seasons.

In a small open sailboat I had taken advantage of its direction and had swooped many miles down the Bay in the lee of the Western Shore. By skimming a short distance beyond the beach my ship was not troubled by the waves which a little farther out were churning the Chesapeake into a mass of whitecaps and streamers of foam. But when late in the afternoon, after sailing some forty miles since dawn, I passed the protection of the sandspit at Point Lookout and plunged into the open reaches of the Bay at the mouth of the Potomac I soon found myself in difficulty. A heavy cross sea, created by one set of waves coming down the Chesapeake and another from the upper stretches of the Potomac, aggravated by a heavy flood tide, kicked up a disturbance which caused my little vessel to careen dizzily as she hit the combers or found them suddenly dropped away beneath. The Potomac was as wild as I have seen it in many years and when late in the afternoon I crossed the twelve miles of river and saw Smith Point ahead I was very weary and glad of a chance to rest.

Among its other peculiarities Smith Point is unique in being one of the few Bay promontories which has a river discharging at its extremity. This estuary, a small forked creek three or four miles in length, is known as the Little Wicomico. Its mouth has been imprisoned between two stone jetties which prevent the channel from silting and becoming too shallow for navigation as it used to do.

Noting the jetties, and being ignorant of the heavy tide that pours out, I sailed blithely into the opening. To my surprise, although the wind was still blowing strong and the sails were straining, the ship barely moved. For a second or two she stood still, held by the surge of ebbing water, then slowly, by inches, proceeded up the racing channel into quiet water beyond. With a sigh of relief I cast anchor in a still pool, tidied up the ship, cooked a much-needed meal, and went to sleep.

During the night, from a grove of pines on shore, a chuck-will's-widow began calling, so loudly and so insistently that I awakened. The out-of-season northwester was still blowing and, although the pool where the boat lay was sheltered, the unceasing rush of air through the needles gave forth a muted roaring much like the turbulence of distant surf. This mingled strangely with the call of the invisible brown bird; the combination was so impelling that I crawled out of my sleeping bag and sat for a while on the deck listening. It was quite dark and, although the stars glittered and gave sufficient light to outline the trees and the more solid objects, there was little to see. Sound and not sight was the dominant and useful sense.

And I was pleased that it was so. It is well at times to forget the visual and to rely on the lesser and more neglected perceptions. In the full glare of daylight, sound and smell and touch are overwhelmed and function only in part. The perfume of flowers is obscured by their bright colors; the aroma of dried earth and dead leaves passes unnoticed; the sounds of birds and insects, although present in abundance, is diminished and lost among a wealth of light and shape, of hue and form. Only at night when vision is gone, when color has become monotone, when shapes are indistinct and ill defined, do the voices, the single notes, the scratchings and raspings, the low whistles, the delicate rustle of grass blades and leaves come into their own and reveal themselves.

Thus, I sat still and listened to the sounds of Smith Point. These were of several orders. Foremost were those of the wind and the sea; they were distinguished by their unceasing character. Close at hand was the diminutive lisping of little ripples gliding up on the black sands of an invisible sand bar, a sort of light accompaniment, muted grace notes to the ponderous bases of the big waves breaking on the beach on the Potomac side of the point. The breakers fell steadily on the sands, in unceasing reiteration. The combers must have been striking the beach at an oblique angle, for the roar of the crumbling water did not begin and cease simultaneously over the length of beach. Instead, the sound had its beginning far up the shore; then rapidly it swept in a crescendo toward the jetties and there terminated in a medley of booming and splashing. No sooner did the sound at the jetty reach its climax than a new wave could be heard advancing and, beyond, the distant sound of another.

But although the surf set the tempo of the water sounds and the ripples supplied a liquid obbligato, the most impelling of all was least obvious. It was a peculiar vibrating hum, a thrumming noise oddly mixed with an occasional hollow gurgle or lisping splash. At first I could not place it; it was close at hand, yet so soft and elusive that at times I wondered if I had heard it. Then, after some moments, I casually laid my hand on the tiller and found the clue. It was the quivering of the rudder and the anchor rope and the hanging centerboard set to vibrating by the flowing tide. In the dark the forces of the sea were steadily at work, the big waves, the lesser ripples; in the depths, the running tide was tearing and gnawing at the sand and the silt.

With the exception of the tide sound, the other noises were wind-created and it was the wind that dominated the night. Its pressure was indicated on all sides, in the roar of the pine needles, the sibilant swishing of beach grass, the rasping of a dead, hanging limb against an invisible tree trunk, in the slap of a loose halyard on the mast, the singing of air through the taut rigging high above the deck.

Early in the gray morning, when I awoke, these same sounds were filling the air. But, in addition, a large number of new ones had come into existence. Half drowsing in that semiconscious interval between sleep and awakening, I heard a multitude of high-pitched cries, oft repeated, shrill and plaintive, vaguely reminiscent of the calls of killdeer, only not so rapid. Mingled with these was the unmistakable clamor of laughing gulls. They seemed to be excited, in turmoil over some avian event, the discovery of a school of delectable minnows or other provender. Other bird voices came across the water, the squawk of little green herons, the out-of-tune cawing of fish crows, the light plaint of sandpipers, and on shore the sweet, liquid, clear calls of some sparrows.

These were the sounds of Smith Point in the early morning—those of the wind, the sea, and birds—and though I did not know it at the time they were the keys to the special personality of the Point and to the significant and unrelated characteristics that cause it to be set aside as unique and different from any other spot in the Bay country.

It was not until after a belated and lazy breakfast that I stepped ashore on the south side of the river mouth and became fully aware of the interesting nature of the place. Close to the end of the jetty was an area in which all the trees were dead. Their whitened, broken branches stood stark against the sky and their fractured limbs littered the ground in profusion. The roots of these trees were nowhere visible; they were shrouded in a blanket of fine white sand. The cause was obvious. The builders of the jetty had dredged the river channel and cast the accumulated silt into the pines; the smothering sand had killed the trees in one mass.

It was not the trees, however, nor the fact that they were dead that struck the eye. Nearly every fifth or sixth tree was topped by a great bulky structure of piled twigs, branches, seaweed, splintered boards, bark, and other flotsam. Many of the structures were six or seven feet tall and half as broad. They were the nests of ospreys and in less than the space of two acres I counted seventeen nests. Beyond were the shapes of many more. There are a lot of places on the Bay where groups of ospreys band together in breeding season, but nowhere had I encountered anything like this.

Circling in the sky above the dead wood were the owners of the nests. They were greatly excited and the air was filled with their whistling. Nearly every step I took caused them to burst forth in new crescendoes of calling and as I passed from nest to nest the birds trebled their voices and came swooping down to shrill close above my head. A few individuals, braver than most, approached so close that I could almost have touched them.

The reason for their excitement soon became evident. When I climbed a whitened and gnarled old conifer and looked down into the neighboring nests, they were nearly all filled with little brown and gray blobs of woolly fluff, young birds not long hatched. In some nests the eggs were still unbroken and they gleamed in sharp contrast to the dark sea grass with which most of the nests were lined. It was the height of the breeding season.

I spent some time in the dead grove quietly seated on the sand, observing the fish hawks and their comings and goings, and I had by that time made up my mind that Smith Point was an unusual spot. But when after a while I got up and idly wandered toward some green pines to the south I began to realize that the dead trees were not the culmination of the wonders of Smith Point but only a suggestion of what was to come.

Every once in a while during the course of a lifetime one chances upon some spot which is superior to all other places, which holds some particular quality, some matchless brand of beauty, or combination of physical loveliness or natural magnificence which is found nowhere else. I can think of several such incomparable places—a certain valley among the hundred-foot sand dunes of wind-swept Assoteague Island in Virginia on the Atlantic coast, of a glade in the Great Dismal Swamp in North Carolina where the ferns grow seven feet tall, of a moss-festooned red sandstone cliff along the edge of the Susquehanna River in Pennsylvania, of a black stream in Georgia overhung with enormous cypress draped with gray-green Spanish moss and of the little-known hemlock-studded canyon of the Cacapon River in West Virginia.

That part of Smith Point to the south of the stricken forest was one of these. When I shouldered my way through bordering vegetation I found myself in a place which seemed to have been lifted entire from some exquisite Chinese tapestry. This feeling was imparted by the presence of numbers of beautiful old pine trees of the long-needled type seen so commonly on old porcelains and on the lacquered screens of a civilization much older than our own.

In few other places have I seen such trees. They did not grow upward in the manner of normal pines seen elsewhere in the Bay country, with long straight boles thrusting toward the sky. Instead, they crouched close to the earth and expended their growth horizontally; their massive limbs hung low and spread out in green-fingered masses; many of the needles ended just an inch or so above the soil. In places little circles inscribed in the sand showed where the swaying of the limbs had left the marks of their motion, gently pushing the grains aside.

These trees had suffered much. Their limbs were gnarled and they were bent and bowed as though burdened with the weight of years. Their tortured branches bore the marks and scars of a hundred winter gales, the crush of summer heat. Ice had hung heavily from the twigs, bearing them down; drifting sand had scoured the trunks; even the Bay had warred with them, for strewn between the trunks were some old bleached boards, spars and logs cast up by long-forgotten storms and laid down to whiten in the sun and crumble slowly to powdery dust.

They were not crowded, these trees; each stood a little distance from its neighbor; the spaces between were composed of alternate patches of brown fallen needles and clean sand. Alone and uncrowded each individual was free to spread as it willed. The result was a most strange parklike effect. There was no tangled undervegetation, no young trees clustering about the roots of their parents. Life was too difficult; the seedlings perished as soon as they were rooted, overwhelmed by hard circumstance; only the strong survived. The glades between the trees were spotless; each brown needle lay precisely where it had fallen seemingly untouched by the wind that had left its mark on all other objects; the sand was as smooth and unmarred as though swept with painstaking care. No one—or so it seemed—had ever been there to mar the scene; no blemishes or unnatural scars ruined the artistry of the wind-sculptured pines or sullied the sands.

This portion of Smith Point was really a peninsula within a peninsula. To the east it was bounded by one of the most magnificent beaches on the Bay, a broad stretch of golden sand; on the west the green twisting channels of the Little Wicomico hemmed it in, securing it against ready intrusion. Only from the south was it possible to approach the area without a boat and dryshod. Thus by natural circumstance has Smith Point remained isolated and undisturbed and uninhabited except for the birds that find it a haven of refuge.

Down the narrow corridor between river and sea I wandered entranced by a succession of matchless glades and beautifully shaped conifers. And all during this walk, for the full mile and a half, until the peninsula terminated abruptly in a tiny cattail-bordered swamp I was the center of a whistling, wheeling swarm of excited fish hawks. It seemed as though all the ospreys of the Bay had congregated there and were filling the sky with their forms.

Smith Point is, indeed, the osprey center of the Chesapeake. The collection of nests that began at the dead forest extended without a break all the way down the peninsula. Nearly every one was occupied; many were located only a few feet above the ground. At one nest near the beach I was able to walk up and, without stretching, lift a young hawk out of its shallow grass-lined depression.

In all, there must be well over one hundred nests in the vicinity. Some were magnificent structures and have been added to year after year. One monster in a low tree near the beach was over nine feet tall and about six or seven in diameter. Enough material had fallen from it to make several nests. Other nests were relatively small and showed evidences of having been just built. One prize structure was built on the beach. It was a considerable pile. The top was about four feet from the ground and it contained three speckled creamy-white eggs.

The variety of material used by ospreys in their nests

is surprising. Favorite items are old fishnets, blackened with tar and whitened with barnacles, odd lengths and sizes of rope, sea grass, boards and planks from old boxes —one such had traveled a long way, for it bore the trademark of a Portuguese wine merchant—the whitened vertebrae of large fish, fragments of old cloth including in one instance an entire potato sack, the rectangular cork blocks of life preservers, pieces of old boats, cornstalks, long lengths of vines, marsh grass, bark, bones, the feathers and wings of dead birds. Among the accumulation was the fabric of a smashed airplane wing, some bamboo presumably lost from a passing ship, cattail stalks, a broken canoe paddle, and a mop dropped overboard by a careless, or exasperated, sailor.

The number of nests was exceeded only by the number of birds. Many were flying, others were sitting impassive on the branches near the nests, resting or adjusting their feathers. Those closest were wheeling and circling in excitement; the more distant were going stolidly about their normal business of catching fish and supplying their young with food. Out in the Bay the stakes of the fishnets were crowded with ospreys, all except one line of poles which had been taken over by a group of cormorants. These, apparently, had refused to budge or had chased the original occupants away.

Among birds the ospreys occupy a unique position. Their habits, their mode of living, their nesting, their temperament differ from those of all other hawks. Alone among the predators, they have attained a universal tolerance from man. In the Chesapeake, as elsewhere, they are accepted as proper and reasonable, as part of the scenery, as a portion of the natural background. Even among the fishermen who pay each year a small tribute from their nets, they are regarded with equanimity and friendly interest. I have never met a fisherman who would think of harming, much less shooting, an osprey.

This discussion of ospreys may seem a divergence from the main theme of Smith Point. But it is not, for the two are inseparable. The hawks are as much a part of the Point as the pines or the wide, white beach. Smith Point is their chosen spot, the hub of their existence; no conception of the region is complete without its accompanying mind picture of the spreading wings, the white and brown feathers, the shrill cries, or the plunging forms of ospreys diving for fish. Smith Point and ospreys are synonymous.

Only those who have witnessed the fishing of these birds can fully appreciate the splendid sight they make and the skill with which they seek their prey. The scene I watched one morning was typical. I was resting on a patch of smooth brown needles, idly contemplating the Bay and some distant ships headed for Norfolk and the open sea. The sky and water were partly framed in an arch of long green needles; the Bay was sparkling and dimpling from a newly risen breeze; the beach gleamed, broad and white; small waves purled against the edge, lisping as they slid against the sand grains.

High in the sky were two birds, both soaring; one was very high, a mere mote in the blue; the other was circling about two hundred feet above the shallows. The sun shone on its belly feathers; they flashed and dimmed as the bird faced or turned away from the light. Once it ceased wheeling and for a full thirty seconds hung motionless, then fluttered and circled again. When it approached the same spot, it halted once more and for a long time hovered, beating its wings rapidly, backwinding to maintain its position, preparing for the strike.

Suddenly the wings whipped back, released their supporting air, and the brown and white body shot downward. Without a halt or pause the bird hurtled toward the water; at the last moment the great recurved claws stretched out and with a crash the osprey hit the surface. There was a flash of spray, a resounding splash, and the bird disappeared.

In a second it was out again. With great heaves of its wings it cleared the surface, shaking the water from its body. Between the claws was a good-sized fish—an adult croaker measuring at least fourteen inches. The fish was flapping and struggling to get free; its tail was quivering and vibrating. I could even see its ridiculous mouth opening and closing at it gasped the unaccustomed air. For a time the osprey had hard going; its load was considerable and the feathers were still laden with moisture. Once it began to lose altitude but the wings beat more violently and it rose another sixty or seventy feet. Then safely above the water it gave a violent shake and I could see the water fall from the feathers. Free from this weight, it gained altitude rapidly and mounted another fifty feet. The fish was still struggling but it had slackened its efforts. The strong sharp claws, one before the other, held it too securely. Through the binoculars I could see a trickle of blood oozing down the silvery, scaly back.

The osprey reached the zenith of its climb and began flying to its nest. It had gone barely fifty feet when it suddenly dived again, emitting a shrill cry as it fell. Close behind was the body of another, much larger, bird. It had come literally out of the blue and it missed the osprey by inches. As it hurtled past, it wheeled and came up underneath, flashing white from its tail and head. It was an adult bald eagle. The eagle turned half over, hung motionless for a second and then dived. As it plummeted down a glistening object fell just in front of it. Before the croaker hit the water the eagle seized it, turned, and went flapping up the Potomac.

The osprey flew inland and sat on a dead branch, where for a brief time it rested as though sulking. But it did not wait long, for shortly after I saw it dive again and come up with a large eel. This time it was not molested, but flew to its nest to deposit the wriggling victim as an offering to its young.

The fish hawk is a specialist, an avian engine designed solely for the catching of fish. There is little evidence of its eating any other food, except perhaps in extreme hunger or under the stress of unusual conditions. Within its normal diet, however, its menu is exceedingly varied. Alewives, rockfish, flying fish, blowfish, eels, pickerel, shad, sunfish, and the ugly toadfish are all fair game. Even flounders are not exempt although how the bird manages to seize and fly away with these flat slippery dinner plate-shaped fish is one of the minor mysteries. That they can see them at all is evidence of keenness of sight of a very high order. Flounders lie flat on the bottom, half covered with silt; their hue fits the pattern of the sand.

Unlike the eagles, the fish hawks do not stoop to eat carrion, nor will they touch tainted flesh. They are hunters of the living and prefer to catch their prey in a clean sportsmanlike way.

They proceed about their business methodically and successfully. After the plunging dive the fish is turned around head first and then, if there are young to be fed, a quick trip is made to the nest. If the fish is small it may be held in one claw; if large, in both. Just before landing the bird releases the hinder claw so that it may alight. The nest is always approached from the same direction and if the osprey is busy with its fishing it may stop only long enough to release its load and then go out over the Bay again.

In one nest on Smith Point I counted twenty-four sunfish laid out in three gleaming rows on the edge of a platform. Each was placed neatly facing the same direction and equally spaced as though some storekeeper had prepared them for sale. The young in this particular nest were too small to feed themselves and I assume the parents were storing the provender for a late dinner, although for the life of me I could not see how two adults and three baby ospreys were going to get outside of twenty-four sunfish, not to mention the additions to the collection made later in the day. However, the next morning none of the sunfish were there. I cannot say for certain they were eaten and not disposed of in some other way, but in any event early the next day the same bird was seen bringing in a two- or three-pound croaker.

Ospreys are models of industry and social behavior. They seem to get along well together and I have never heard of their quarreling with one another in the raucous manner of crows or stealing one another's food as the jays do. Their colonies appear peaceful and well ordered. Their relationships with other birds are excellent. Ospreys mind their own business strictly and seem willing to live and let live. Without quibbling they will share their ponderous nests with grackles, starlings, and any other small fry which wish to make use of the crevices and interstices of the lower floors, so to speak. In one nest on Smith Point no less than seven pairs of grackles had spaces on the lower levels and in a couple of the very low nests the meadow mice had established themselves in numbers.

Indeed, I am prone to regard the fish hawks of Smith Point with admiration. In their well-ordered way they live out their independent yet sociable lives. Their sustenance is free and provided with a prodigal hand. With the exception of an occasional, and harmless, brush with an eagle they pass their days without undue frustration or inordinate ambition. Beneath their soaring wings is spread the whole vast expanse of the Bay to roam and to feed from. There is no one to say nay to their avian pleasures and they have earned the respect of their kind and of man. Only the elements—the sun, the wind, the rain, and an occasional storm—offer resistance. Theirs is an honorable station; their part in life is filled expertly, efficiently, and with dignity.

CHAPTER 10

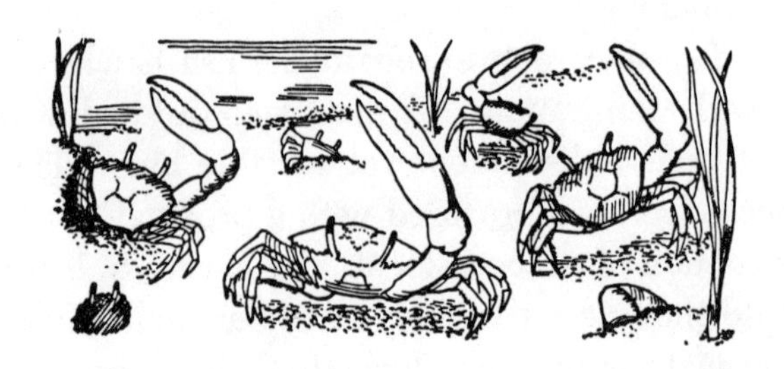

THE SIGN OF CANCER

For the space of a half hour between seven and eight o'clock one summer evening I lay on the sand near the mouth of the Rappahannock River in Virginia and tried to imagine what it was like to be a Chesapeake Bay fiddler crab. This proved to be a highly unscientific, wholly undignified, yet entertaining proceeding.

In order to secure the proper perspective it was necessary to lie prone, half buried between land and water, and so adjusted that my eyes were but an inch or so above sea level. At this height the world took on new dimensions and assumed in the half-light of evening an aspect so unreal as to appear slightly Martian.

It is a revealing commentary on the steadfast habits of the human race that not one person in ten thousand ever thinks of looking at his surroundings except from the long-established adult elevation of between five and six feet. Within this narow zone our visual perceptions are limited except for that transient period of childhood when the family dining room is a large space interspersed with a multitude of wooden legs supporting a variety of tables, chairs, and other mysterious articles possessing unpolished and undecorated undersides.

Viewpoint is a neglected subject whether applied to natural objects or such abstruse subjects as philosophy and religion. The earth in the eyes of an astronomer is a mere speck in the evening sky, a microcosm lost in space. But it is also a vast globe twenty-four thousand miles in circumference strewn with leagues upon leagues of empty ocean and endless stretches of forest, prairie, mountain, or desert. Viewpoint is the difference between the statement that religion is the bulwark of the people or that it is the opium of the masses; that Jesus is the Son of God or only a Galilean thaumaturgist. It is the reason why the world has so often seen the spectacle of opposing armies moving into battle with the same rallying cry resounding from both ranks. Viewpoint is the narrow space between illusion and disillusion, between tears and laughter, faith and cynicism. To a girl of six a broken doll is an irreparable tragedy; the loss of the same toy at sixteen is a matter of indifference.

A Chesapeake beach is not the same place at six feet, six thousand feet, or at one-sixteenth of an inch. From a plane it is a narrow curving ribbon wedged tightly between masses of green. At man height it is a pleasant place to go swimming, a region of smooth soft sand and littered seashells. From an ant's perspective it is a terrifying desert of jagged, multicolored boulders, a Sahara of frightful hills and deep valleys shorn of any traces of green. It is an endless dangerous place of crumbling loose rocks, of avalanches, and of suffocating heat. It is a place where when the wind blows the air is filled with great sharp stones which bound across the valleys and pile up in long glistening mountains.

The spot where I began my unconventional investigation into the viewpoint of fiddler crabs was where three distinct places met in neat juxtaposition. A diminutive patch of white sand, a considerable region of grassy mud-filled marsh, and a long glistening mud bar met in a triangle in a little indentation at the mouth of the river. Not too far away a deep salty channel bordered the bar and carried the tide to and from the Bay. The tide was still ebbing and the bar was momentarily growing larger, creeping gradually into the open air.

In the long rays of the setting sun the bar gleamed red except in spots where the tide had left long ripples undulating across the silt. Between the ridges the depressions were steeped in dark-maroon shadow which gave them an exaggerated effect. From the height of one inch they appeared as long low mountains strangely drenched with half-dried blood.

But it was not the bar or the beach that provided the setting for this fiddler crab landscape. It was the marsh itself, the dank region of the swamp. There were tumbled masses of black mud fissured by deep ravines and canyons; the sides of these were riddled with hundreds of black holes and jagged-mouthed caves. About the mouths of the caves the moisture left by the tide drooled downward leaving little trickles of dark slime. Between the caves a tangle of matted and twisted roots trailed fantastically over the ravine faces; high above, a forest of grasses waved and nodded in the wind.

For a long time there was no movement except for the swaying grasses and the gentle drip of water along the canyon faces. Like the wet on the sand bar, the ooze of the canyon walls and the ravines reflected red, crimson in the high lights, maroon and deep purple in the distance. As I watched, the lower edge of the red crept upward, a fraction of an inch at a time, leaving the depths in shadow.

Then, when the last crimson gleam had gone and the sodden damp earth had become gray and brown and black, when I had been motionless for long enough to be accepted as part of the soil, the fiddler crabs came from their shelters in the wet earth. I have seen fiddler crabs many dozens of times but never until that evening had I realized what strange beings they really are.

They were preceded by a queer rustling noise, reminiscent of the sound the bubbles in the last wash of surf make as they are breaking, a peculiar crackling sound, but on a very small scale. This rustling was accompanied by a multitude of faint scratchings and scrapings and tiny wet sucking sounds.

Suddenly there they were. From each cave and crevice there protruded pairs of glistening eyes which surveyed the glooming world carefully. The eyes were thrust out on short bulbous stalks and they swiveled from side to side searching for possible danger. Then slowly, with many false starts and retreatings, the owners of the protruding eyes inched out of their burrows and crept into the sodden ravines.

From any normal position the sight would have been

average enough, but at sea level the sudden peopling of the dark marsh with hundreds of outlandish crabs, each carrying an enormous claw, gave the view Dantesque proportions; imparted the feeling that this was not a scene on the Chesapeake Bay in Virginia but a summoning of otherworld beings on Jupiter, Mercury, or Mars.

Imagine that owing to some dread disease, glandular maladjustment, or some other horrible cause your hand should suddenly begin growing, day after day increasing in size, gaining steadily in length and thickness until it was all of eight feet long and weighed almost a hundred pounds. And that in this state you would have to carry it forever waving in front of you, a perpetual impediment. Then you will have some idea of the problems of fiddler crab life. For one claw of each male fiddler crab is thus extended, although in proper proportion.

And so, in the dark, it appeared as though the marsh had suddenly become alive with great curving ivory claws being inexplicably carried about. So disproportionate were these attachments that their bearers were eclipsed, half hidden behind the monstrous appendages.

The purpose of this curious development is not entirely clear. Only the males are so afflicted and only in one claw. The claws are seldom used for any purpose except for signaling or intimidation. As weapons they appear to have no real value; the life of a fiddler crab is, in reality, one great fright after another. For all their superior armament they are among the most timid of the marsh dwellers. The slightest movement startles them into a frenzy of panic and if, during the course of their feeding, they stray far from their holes and are frightened or become suspicious, they flee pell-mell for shelter. Waves of panic engulf them and if one member makes a dash for home he is usually followed by the whole frantic horde. And then, unless long peaceful minutes pass, they will remain quiet in their burrows, hidden until all hint of danger is gone.

To see one of these fiddler panics is ludicrous. On one occasion my body was between the crabs and safety and they poured over my legs in their haste to seek sanctuary. Although there is not a sound to be heard, so violent are their exertions that one can almost imagine them rushing down the beach screaming.

Actually, however, the claws appear to have other purposes than fighting although I have noticed when two males meet or when one encroaches on the territory of the other a great deal of brandishing on both sides is resorted to. The males rush up to each other, wave their claws, opening and shutting them, retreat and rush up again. Sometimes they lock jaws in futile, harmless contest; for minutes there will be much pushing and heaving; but no blood is ever shed and sooner or later one decides it has had enough and retires from the field leaving the other triumphantly waving. Somehow, whenever this occurred I was reminded of the referee holding up the victor's hand after a prizefight in Madison Square Garden.

The amount of claw waving between disagreeing males is as nothing to that which marks the progress of a female through a group of male fiddler crabs. As she passes the males rush up to her gesturing frantically until she is out of sight. It is a fiddler crab's bid for attention and, in amatory intent, probably differs little from the whistles that proceed from the drugstore corner at the passing of every likely-looking prospect.

Beyond its virtue as an attraction for fiddler females the possession of a big claw is distinctly a disadvantage.

Fiddlers feed on small matter cast up by the tide and they like to glean from the newly formed windrows of mud and sand washed up by the ripples. This they scoop up and ladle into their mouths. The small claws are admirably shaped for this purpose and it would seem that the females, possessing two scoops, are better fitted to survive than the males. Indeed, a male which loses its small claw is likely to starve to death before a new one grows even though an abundance of food is within easy reach.

On an evening in early spring, shortly after the warm sun had tempered the chill of the water and heated the black soil of the marsh, at the hour of dusk, on an evening just like the one when I lay at fiddler crab height on the sand, a female fiddler crab in company with a host of her sisters, crept down to the water's edge and stepped warily into the Bay. This in itself was unusual; although the fiddlers are true crabs and, as such, creatures of the sea, they avoid their mother ocean as much as is consistent with their habits of gleaning from the very water's edge. It is all very well for a fiddler to venture to the last swirl of the incoming ripples, or even to get the tips of its legs wet, but to venture beyond except for reasons of extreme urgency is to court instant and certain disaster.

About the borders of every fiddler colony is lurking an ever-present horde of hungry carnivores, long needle-toothed ravenous fishes, waiting for some luckless fiddler to fall overboard accidentally or injudiciously stray too far. Thus it is that with the rise of every tide the fiddlers flee slowly before it, falling back to their holes and shelters. And when the wet flood allows them no further retreat except to dry land where other legions of hungry beings are waiting, they go into their burrows, seal up the entrances, or allow the wave-washed sand to fall in above them. There in the dark, covered by the wet soil and salty water, they wait immobile, resting or dreaming whatever dreams fiddler crabs are prone to, until once again the sea releases them for further activity.

But on this evening there was real need and a good reason to brave the open waters. For beneath the body of each female, attached to the plastron, were bunches of tiny fertilized eggs. It was time for the beginning of the cycle of reproduction, time for the return of the young to the sea, the annual pouring out of new life into the waters.

In the half-dark the females ventured out until all of a half inch of rushing water swept between their legs. There the young were cast forth. With swift violent flicks the newly born fiddlers were jerked from their cases and hurled into the black flood. As seed is sown in the field so were these fiddlers-to-be cast upon the waters. From each mother's body went several hundred; the Bay was filled for yards about with their twitching bodies along with a host of other microscopic beings.

It is a wise fiddler crab that would know its own progeny, for the newly hatched young are as unlike the parents as it is possible to be. If a human being were to give birth to a mosquito or an elephant to a flea or should a violet suddenly begin sprouting hollyhocks, the comparison would not be exaggerated. Baby fiddler crabs, newly born, are as weird little beings as were ever conjured out of a fevered imagination.

They are glassy and, except for two monstrous blackish eyes, are as transparent as the water itself; their bodies consist chiefly of a pair of needle-sharp spines. On either side of the misshapen head and thorax are tufts of feathery hairs which serve as oars of sorts and which propel the unbelievable creature through the salty water at a dizzy, erratic pace.

The first hour of a fiddler's existence is a hazardous one. Every inch of the warm Bay water swarms with a hundred greedy animals. There are hungry larval fish, innumerable stinging jellyfish, minute crustaceans with snapping jaws, cruel hooks, barbs and spines, with nets and snares all waiting for prey. And then, as if these were not enough, there is the ever-present danger of becoming entangled in the filamentous strands of seaweed and algae to perish miserably, locked tight in an underwater prison. Each ripple may be the one which casts the young on the beach to dry on the hot bare sand—or, even worse, to be scooped up by one's own parents and devoured. Countless numbers perish in this manner, their lives blotted out before they are fairly begun.

But if by some fortuituous combination of circumstances the zoea, for so the tiny being is known, survives the first hours without being swept helplessly out to sea or buried in the mud or disappears down the gullet of some creature more misshapen or weird than itself, it spends its time for the next several weeks struggling and fighting toward the surface. Its aim in life is to keep as close to the top as is possible and there, in its frantic way, to scull itself along, grasping and devouring other creatures smaller or less swift than itself. During these weeks its appetite is astounding and it clutches everything within reach, eating and in its turn trying to keep from being eaten.

As the days pass, it grows slowly and molts, changing slightly with the donning of each new skin. And with each molt the difficulty of keeping at the top of the water becomes greater until, at last, with its sculling apparatus blunted and useless, it sinks into the green depths. There for a space it wallows helplessly in the mud and silt.

With the sinking into the mud its appearance does not improve nor does its nature. Like some ugly-dispositioned dragon, it crawls, groveling, along the bottom. It snaps irritably at everything it touches, including its own brothers and sisters, and woe betide any younger relative it may encounter. But the beast is no longer a zoea, for with the last molt before sinking to the bottom it casts aside forever its zoeal shape and becomes like a crab, although a more ugly, ill-proportioned crab would be difficult to imagine.

In this form it is all of a sixteenth of an inch in length and for the next several weeks spends its hours alternately clinging to bits of weeds, sticks, or other debris which may come within its reach. And then, if fate is kind, it will someday find itself drifted ashore, perhaps at the very spot of its nativity, perhaps many miles away. As like as not it will find itself faced with the necessity of running a gantlet of its immediate ancestors and some distant relatives which, without the slightest compunction or tremor of parental concern, will snatch it up and crush it between strong hard jaws.

But once past this barrier and beyond the reach of the waves it digs a diminutive burrow and retires temporarily from its labors. It has well earned a rest. For all its short and hazardous life it has never had a moment free from the threat of accident or hungry attack. Thousands of its brothers have perished hopelessly, lost in a waste of water or cast loose on an unfriendly beach. But at last the crab has a haven of retreat, a castle and a fortress.

And although this may be built upon the shifting sands, it is secure from the world.

From that time forth the young fiddler is seldom seen far from its burrow, and it is to the burrow that it retreats on the slightest suspicion. Storm and high tide, gale or hurricane, down in the wet depths the fiddler crabs know, by an instinct which never fails, that in time all will be well; that in a few hours, a few days, the tide will return to normal, the winds will cease howling, and once again the sun will shine on the exposed sand bars and mud flats and the little ripples will again deposit the stomach-filling sea substance on the beach in endless quantity to be scooped at will.

The life of a fiddler is oriented about its home. As it grows, molting time and again, the tunnel is enlarged or perhaps a new one is started. But new or old the burrow is the focal point of the life of the individual; as its brothers and sisters in their turn struggle out of the surf the numbers of burrows increase until, in many cases, colonies of many thousands are concentrated in one spot. And although the fiddlers never henceforth leave their homes, their race has traveled far; their kind are found all over the lower portion of the Bay and up and down the coast wherever conditions of earth and tide and sea are suitable.

Although fiddler crabs think no thoughts, it is not unreasonable to assume that each crab has the same basic regard for its burrow as a human being for its home. It is certain that each crab recognizes its own tunnel and will retire to it even when other burrows are closer. How they distinguish their holes from those of a thousand others in the same vicinity is not clear. A full-grown fiddler colony is a maze of little openings all looking as much alike as the rows of houses in human cities—and as devoid of individuality. But where houses have numbers and other recognition marks, the holes of fiddler crabs have no identification, recognizable at least to human eyes. Yet once when I marked a number of captured fiddler crabs and moved them some yards away from their tunnels, they found their way unerringly home again.

In the mind of a fiddler crab the world is an expanse of sand and mud radiating from a hole in the ground. The tide falls so that the crab may come forth, or it rises so that the drying walls may be moistened again. The waves come in a certain distance so that one need not go too far away to feed and so that one can get home again with dispatch. The sun moves to illuminate the entrance and the winds blow to ventilate it. Life is predicated on the happenings within a clearly defined circle; existence is not linear for a fiddler crab; it is radiate, a revolving about a hub; it is a 360-degree and a 10-foot limitation. Adult fiddler crabs are bound to a literal wheel of life; the spokes are the continual perambulations in search of food or a mate; the rim is as exact as if drawn with a compass; passage beyond the periphery is unthinkable; outside the world does not exist nor is it dreamed of. A fiddler colony can be conceived as a congregation of overlapping circles, the burrow of a crab the center of each.

This is an almost human arrangement, for most of us are bound similarly to a circle of existence. There is probably little fundamental difference between the fiddler crab that travels daily from its burrow to its favorite feeding place and the man who climbs into his car or boards the commuter train for a day at the office. The purpose is the same.

However, the life of a fiddler colony is human-like only

in its superficial aspects. Fiddlers possess houses, live sedate and well-regulated lives, congregate in cities and villages. But they are strangely solitary for all their seeming social activities. They act as individuals, prefer to live alone in their burrows, are hermits in the midst of numbers. While they may move or feed together, they do not act in concert nor do they have leaders or commanders; they are perennially social antisocialists. Which may explain why no one has ever observed the residents of a fiddler village marching en masse to impose their will on the denizens of a neighboring town or to solve a dispute over property or eminent domain. Fiddlers are primitives; they possess no brains to speak of, but they seem to solve their peculiar problems with marked success and without apparent dispute. Their small place in life is assured and they are blessed with all the worldly goods their crustacean natures require: a home, food, and a never-failing green sea to replenish their wants forever.

CHAPTER 11

THE BAY IN TIME

It was bitter cold. An icy wind out of the east swept singing across the valley. With it came snow and sleet, freezing on the ground and burying the stunted vegetation in glassy gray-white mounds. The sun had faded days before and a gray gloom hung over the earth.

In the deepest part of the valley a long curving white ribbon marked the course of the great river that was to be known in much later times as the Susquehanna. Miles away lay a similar watercourse, broad and choked with ice floes and small bergs—the Potomac. At their joining place the tortured ice was heaved up in confused ridges, great three-foot thick sheets piled one upon the other in tumbled masses. The ice had come down the river valleys from miles away; and, embedded in the floes were granite boulders, fragments of reddish-brown breccia, and sheets of broken red sandstone. In the last flood in the last thaw before the savage cold had set in they had been ripped from the riverbanks far up in the hills and carried scores of miles down toward the sea.

But the sea was still far away, for the junction of the two rivers was nearly a hundred miles from salt water. Winter after winter the cold had increased, for hundreds of years the summers had grown shorter, the warmth less pronounced, the sun more wan, more frequently obscured by clouds and hovering mist. And in all this time the ocean had steadily retreated, a few yards a season, maybe a fraction of a mile each forty or fifty decades. The amount of water in the ocean was diminishing, evaporating steadily, blowing away in white billowing clouds or in gray overcast of luminous fog. All over the polar regions the condensed water vapor was falling as snow, gently, quietly, piling ever higher. Thousands of square miles of snow crystals and solid ice lay upon the earth, a mile deep in places.

Not too far away, about a hundred miles to the north in a straight line, the edge of the great glacier lay glittering in the wan winter light. Millions of tons of hard ice blanketed the earth; from just above the future site of Philadelphia to the northern pole was an unbroken, jagged, bluish-white sheet. It was the time of the culmination of the great ice age; the earth was frosted and life had retreated to the south, driven on by the cold and the bitter winds.

The ice had come slowly at first and then more rapidly as century succeeded century. But first the land had heaved as though breathing deeply. In northern New

England and as far south as Long Island tall stone cliffs rose up a thousand feet; they were crevassed and seamed with deep dark fjords which penetrated long distances into the interior. In Maryland and Virginia the land also rose, but the swelling was not so great, a hundred or a hundred fifty feet at most. But the heaving of the continent, high in places, less in others, served to increase the land mass, cut off or displaced the warm ocean currents, and began the successive climatic changes which in their turn brought the years of ever-burgeoning cold. And as the land tilted and thrust upward, the waters of the Bay, of the prehistoric Susquehanna retreated and slipped back toward the borders of the continental shelf.

It was the beginning of the end of the great age of mammals; the culmination of the period that produced the horde of strange and mighty creatures, the titanotheres, the great sloths, the giant plated armadillos, the brontotheres and all the varied beasts that had their beginnings at the close of the age of reptiles. They met their nemesis individually or collectively, not because they were insufficiently fleet of foot, or strong, but because the earth heaved, the ocean currents were shut off or shifted, because the sun no longer shone as before and was replaced by interminable rain and clouds, because the rain turned to snow, the snow to ice, because the vegetation and flowers vanished, flooded or frozen, because food failed or went elsewhere. When the forests disappeared the forest dwellers moved on; when the meadows became acres of slush the grass eaters drifted to other places. For days and weeks or years the elimination of life went on, slowly or rapidly. Thousands of beasts, the preyed-upon and the preying fled before circumstance.

Down through unmarked Pennsylvania and along the Atlantic coast the hordes of displaced beings moved, urged ever onward by altering conditions, stopping at the deep valley that was the Bay only long enough to find a way across.

For innumerable months and years the procession had continued; countless hordes had crossed, fording the water in shallow places in the upper Bay near the fall line where the Susquehanna dropped in rapids and innumerable riffles from the Piedmont Plateau, from rock to rock and boulder to boulder. Others crept across the winter's ice, each season growing thicker and more solid. Some few swam, others failed and stayed behind. These perished miserably, shivering until death overtook them in numbing sleep. The trees and plants rooted to the soil simply died and were covered, at first the delicate and fragile, then the most sturdy. Only the mosses and the lichens persisted seemingly unaffected.

At first the migration was slow, but then it gathered in volume as the beasts were driven out of the north. Great shaggy mastodons lumbered across the river, breasting the water, trumpeting as they moved, the steam from their breath showing plainly in the air. Their time was nearly gone, they were moving to extinction, to join the long line of other beings overwhelmed by time and adverse circumstance, to become brothers to the dinosaurs, the pterodactyls, and other lost creatures. With them went bristled, tusked peccaries, camels with diminutive humps and supercilious noses, advancing to oblivion.

Others made the crossing and lived to move on. Bear, elk, and deer struggled over the barrier and large gray wolves kept them company. The retreat was shared by all, the carnivores, the vegetarians, and the insectivores.

Even the fishes sought new abodes, moved to the eastward down the rivers to the blue salt ocean. As the Bay grew smaller, more narrow, and more fresh, the sea creatures abandoned their haunts or limited their accustomed migrations.

There is little left to tell of these times. The bones of the overwhelmed have been ground up or washed away, cracked and split by ice, crumbled by the sun. Here and there in isolated localities a few pathetic fragments remain. A few teeth of the mammoths and mastodons have been found, the partial skeletons of some camels, tapirs, and peccaries, of wolves, bears, sloths, and porcupines. In the immediate region of the Bay the remains are scarce, but from the paleontological evidence of nearby places in Pennsylvania, New Jersey, and Virginia we know that a large assemblage of animals made the Chesapeake country their home. Saber-toothed tigers roamed the meadows, prowled the river valleys and the beaches, and together with jaguars preyed upon the herbivores. Huge bears, as big as or bigger than the modern grizzlies, climbed the hills, undoubtedly robbed the bees of honey as bears do even unto this day and perhaps caught the shad in spawning time as they ran up the rivers to lay their eggs in fresh water.

The broad outlines of the events of those times are discernible; we know that the ancient animals existed because their bones have been found and interred with proper scientific reverence in a score of museums. The observant eye can still see the evidence of the heavings and settlings of the land. All over the tidewater region are the remnants of ancient shore levels, sandy and gravelly terraces marking successive changes in elevation. Capitol Hill in Washington is one of these. But of the struggles of the individual beings, of their comings and goings, of their efforts against the elements there is little trace. No accurate description will ever be written, for none is possible; as in the preceding paragraphs, the only refuge of the writer who would tell of those times is conjecture, speculation based on slim evidence.

Yet no consideration of the Chesapeake Bay country is adequate without recognition of that which has gone before. Geography, like the mathematics of Einstein, should be a science of four dimensions. The factor of time cannot be ignored; it is expressed everywhere, the very rocks and boulders testify to it, the sand is saturated with it; time broods over the creeks and in the valleys.

Time is of the essence. Just as the music of Johann Strauss has added meaning because there once existed a gay and joyous Vienna and just as the melodies of Schubert are haunting and have sad undertones because they echo the long-gone feelings of the man who composed them, so the features, the atmosphere of the Bay country reflects the happenings of the past. The Bay itself exists because ten times ten thousand years ago a river valley was drowned, because it slid slowly into the sea.

On the beach at Gibson Island in Maryland, where the fashionable go bathing, in the sand are tiny yellow specks, clear golden fragments not wholly unlike the millions of other fragments that go to make up the beach. They are fragments of amber, tiny solidified pieces of resin which once oozed from green trees, from the evergreens that carpeted the area for miles. In the warm summer the tiny droplets oozed over the bark, perhaps at some spot on the trunk injured by the rubbing of the antlers of an elk, scratched by a bear to sharpen its claws, or perhaps on a limb where some irritating insect gnawed at the wood. The forests are gone, the elk and the insects, the trees have merged with the elements, the amber remains.

However, only a short distance away, in the banks that line the north shore of the Magothy River are black lines in the yellow clay. These lines are carbonaceous material that was once trees and grasses. Embedded in the clay are wisps of blackened twigs coated with grains of fool's gold, iron pyrites leached from the soil above. And at Pinehurst on the Bay front, protruding from the cliff faces, close to the water and hidden by the bulkheads of the cottages of a modern shore resort are the tangled roots and stumps of what was once a magnificent stand of cypress. Above them rise fifteen or twenty feet of cliff.

The evidence is plain for all to see. The trees grew tall and straight at the level of the sea. The water crept between their roots and knees and lay in quiet black pools. Mosquitoes hummed in the shadows and, no doubt, long streamers of gray Spanish moss were draped in festoons on the branches. In time the land lowered, the water crept up the trunks, the trees died or were smothered in silt, in soft clinging mud which filtered between the roots and bound them fast. The upper portions of the trees fell to earth or dried up and crumbled away. And all the while the land kept sinking, the silt settling, until the forest was buried beneath tons of sand and clay.

For thousands of years the stumps lay buried. Over them the waters rolled, winds dimpled the surface causing it to ripple and dance. For a long time all was still. Then slowly the land began to heave. The waters drooled away and plants and flowers bloomed on the dry land again. The soil that had accumulated, however, began to wash away. The Bay waves gnawed at the newly exposed shore, ate it away a little at a time and once more revealed the buried stumps. So today they sit, waiting for the long-delayed and this time final destruction.

Time has most prodigally revealed its hand on the western shore of the Chesapeake from just below Chesapeake Beach to Solomons Island in Maryland. There, for nearly thirty miles, is flung a long series of tall cliffs and bluffs, high, steep areas of blue clay and yellow sandy soil falling for over a hundred feet in places sheer into the water. The waves lap unhindered against their base except for a narrow strip of sand which often disappears at high tide.

The cliffs are the locality from which was first described the deposits and fossils of the Miocene period in America, the great middle period of the age of mammals. Strewn singly and in great layers of solid shell, in places twelve or more feet thick, throughout the cliffs are the remains of a once-abundant marine life. Huge extinct scallops more than the span of a hand across litter the clay in profusion, gigantic oysters occur in thick bands where once they lay on the ocean bottom, clamlike mollusks each over twelve inches long occur in frequent clusters. The rib bones of porpoises and giant whales, together with their massive vertebrae, may be picked up by the dozens from the cliff debris at the water's edge. More rarely are found the bones of birds, the carapaces of giant sea turtles, the imprints of fossil crabs. Crocodile teeth and skeletal remains are not infrequent. Sharks also must have lived by the thousands, for their teeth are everywhere; some of these are six inches in length, more are smaller; one can gather hundreds simply by walking the beach below the cliffs and picking them out of the sand. Examination of certain of the clays and fine sands from the cliffs reveals also a large fauna of micro-animals, diatoms, foraminifera, and similar minute beings.

Thus, when the inner portion of the North American continent was undergoing profound changes, probably during a period of great droughts and enormous sandstorms, when Kansas and the midwestern states were a dusty dry desert, the Chesapeake country was a lush steaming region of oceanic lagoons and marshes probably reminiscent of present coastal South America. It is likely there was no Chesapeake Bay at all then, only miles of shallow salty channels, broken by winding sand bars, open to the ocean not too far away. Here the sharks and porpoises came to feed on a teeming fish life, and here the whales were often stranded or drifted ashore if one can judge by the frequency of their unmistakable bones. The bellowing of crocodiles filled the air, turtles crept on the sandy beaches to lay their eggs, and swarms of sea birds swirled excitedly over the bars and tidal flats.

There was, of course, no hint of man; he had not yet appeared on the North American Continent—nor anywhere else in the world, for that matter. Man, the abstract-thinking mammal, had not yet been conceived and it is probable that his distant preanthropoidal ancestors were still struggling for survival in the Asiatic or African jungles.

The touch of the hand of early man as a primitive being is, however, evident everywhere in the Chesapeake. The pre-Columbian Indians have left their trail in many places. Long layers of oyster shell on nearly every river and sizable creek indicate their campsites and once-permanent tidewater villages. There are even a dozen or so spots where along the eroded Bay shore, notably in the Patuxent River valley, one can see the remains of their campfires, oval blackened holes in the ground filled with charcoal, broken bones of deer and smaller mammals, and charred, roasted oyster shells. Some of their shell heaps cover acres and their villages numbered hundreds of individuals. They nearly always selected choice spots, generally on the point of some quiet sheltered creek just off a larger river. Today many a summer cottage is perched, its inhabitants unaware, on the site of a once-prosperous Indian town.

The very names of the rivers are a legacy of a vanished period. Indian names—Wicomico, Rappahannock, Choptank, Pocomoke, Potomac—abound. The meanings are no longer clear but their sound is an echo, a diminishing note from the era of a primeval people. Thus the fancy of some tribal chieftain, the poetic instinct, perhaps, of some aboriginal individual, is projected beyond his time; the namer remains unknown, forgotten or never recognized; but the euphony of his expression is carried forward, just as the melody of a concert persists, stored in the cells of the brain, sometimes for days after the music has ceased.

I recall once standing in front of a pair of opposing mirrors where delineated in myriad repetition were a series of diminishing images, of myself and of the room, each reflection becoming more and more indistinct, less and less clear yet still visible until one had the feeling of gazing into infinity. And so the Chesapeake; it is a figurative mirror in which the nearer events are plain and clear, the older vague and ill defined. Last week's storm throws trees to the ground; their broken limbs and drying leaves are starkly visible; the trees of the storm of a decade ago are only moldering wood, moss-covered mounds on the forest floor; the tree of a thousand years, a carbonaceous streak in a claybank.

The mirrors of the Bay are the beaches and the cliffs, the silt of the ocean floor, even the surrounding vegetation itself. Not long ago I found in a small ravine near Drum Point, Maryland, growing close to the water's edge, a little clump of horsetails, those exotic tubular plants which are almost the sole survivors of the Carboniferous period, the age of the creeping amphibians and of the lush vegetation that produced our present-day coal. It was easy to half close the eyes and allow the vision to distort the plants into a picture of an antediluvian swamp, an imaginary venture into a hundred million years of time. And to show that this is not pure imagination or a literary fiction, only a few miles from the center of Baltimore City, in Westport, a grimy commercial suburb, were once dug up the remains of a whole forest of cycad trees, palmlike plants with trunks resembling nothing so much as gigantic pineapples. They were contemporary with the primal horsetails, when horsetails had trunks a foot in diameter and stalks forty feet high. The cycads are gone, only the dwarf progeny of the horsetails remain, living and relict testimony, a vegetable extension of another time.

And so, with time expressed as a relative concept, the Chesapeake, as every other place on this earth, becomes properly fixed in the evolutionary progression. In geological terms, in its present form, it has been in existence for only the briefest span, at most only a million years or so, probably much less, a mere nothing as time is reckoned. The Bay was young when the ice came out of the north, it is now only in early middle age. It is probable that before its full maturity the hand of the despoiler, the precocious mammal man, who has littered the beaches and befouled the waters, will be gone as if he had never existed. When his bridges and wharves, cities and resorts are crumbling mounds of masonry and reddish rust, the Bay will still be there, the waves will dance as now, the fish will still come in the spring and depart in the fall, the beaches will once more be clean and bright. And before the Bay slides again into the sea to be buried under tons of crushing water, or conversely to be lifted high above its present level, or another time smothered in snow and ice, man and all his works will have become one with the mastodons, the mammoths, the camels, and the vanished legions that were part of its youth.

CHAPTER 12

THE TOP OF THE BAY

THE TOP OF THE BAY, ACCORDing to the atlas, covers about twenty-seven hundred square miles. It is a place of two dimensions—length and breadth—although this point may be disputed by any waterman who has sailed it in a gray northeaster or who has had to tack across it in a summer squall. However, for literary purposes, the vertical may be ignored. The surface is a dividing line, a separation only.

Whenever I contemplate the surface of the Bay, or of water anywhere for that matter, I somehow think of a famous dispute which back in the Middle Ages disrupted the clerical calm of the monasteries and even distracted the more secular activities of the royal courts. The quarrel centered in the amazing question of how many angels could stand on the point of a pin! It went on for years, blood was shed over it and bitter were the feelings it aroused. I believe it was St. Thomas Aquinas who finally brought rationalization into the argument and showed it for the silly thing it was.

Now, this medieval wrangle has no real reference here except as a reminder that the surface of the Bay can be considered from a double viewpoint. The surface is the top of the water—or the bottom of the atmosphere, as you please. Like the angels on the pin, it really doesn't matter; the illustration, however, is helpful in emphasizing that the surface, as such, is not an entity; its thickness is zero; it is lateral extension and nothing more.

All of this is important in securing the proper perspective for an inspection of this familiar scene and an appreciation of its character, commonplace when looked at from the usual viewpoint; unique if considered from unorthodox grounds.

Ordinarily, from the upper air, the surface of the Bay is its most expressive feature. It is seldom twice alike. In calm and in the hush of a summer evening it has a brooding quality, a certain air of somnolence; one is reminded of a slow breathing as if the earth were slumbering. Then there are the days when the Bay is a frothing mass of swirling bubbles, long streamers of white spray with gray-black waves for contrast. Most wonderful are the times of the northwesters, when the sky turns blue and the surface azure in reflection; these are the days of sparkle and sunlight, sailorman's weather. And there are the nights, the hours of black-purple shadows, the reflected gleam of ships' running lights, or the resplendent sheen of the moon in a light breeze.

There are few places on the North American continent where the mood of the water undergoes so many alterations, and in such succession, as in the Chesapeake. Like the emotions of a spoiled woman, they vary from hour to hour. All the common elements of feeling are there, anger, placidity, sparkle, sullenness. And, woman-like, the succession may be swift; brooding may succeed gaiety in a few moments; lustiness may come swiftly after gentleness; conversely, wrath may melt into placid calm. Fickleness is the keynote and, like the telltale signs in the face of a changeable woman, the Bay's mood and the warning of its transformation are revealed by the surface.

Old Bay hands know that a slick blue sea covered with a multitude of soapy bubbles is the prelude to one of the Chesapeake's famous midsummer squalls, that the lack of a cloud on the horizon is a deception and the gentle roll of the waves a delusion.

Although the calm may last for hours, the time comes when a line of shadow rapidly forms in the distance, when a dark arch of clouds extends across the whole sky. The arch may be miles away but with increasing swiftness it spreads; I have seen line squalls, as they are called, cover the whole sky in the time it takes to don a pair of oilskins. Close by, the water is still calm and blue but under the clouds it has lost its clarity; instead, it is gray, almost black, and ahead of the dark is a border of white. The white is almost exactly under the line of clouds, and it moves as they move. And the clouds themselves are now seen to be black angry masses of turbulent vapor. In the front they are churned and roll over and over. The contrast between the blue sky and these inky clouds is startling. Ahead the sun is still shining; behind, like the backdrop of a theater, is a solid mass of luminescent gray; this gray is impenetrable; it is cold and metallic.

For a few moments the heat of the day remains; then a cold air stirs the water, a few raindrops fall, pattering.

The calm descends again, but not for long. The white line is very close and it is leaping forward, the surface is shredded and torn and the line is a mass of low writhing white froth. Close behind comes a driving sheet of rain, flying almost horizontal to the wind. With a roar it hits, and the sky is filled with tortured liquid and screaming air. From dead calm the wind has risen to fifty miles an hour in the space of thirty seconds; sky, land, and sea are blotted out.

The first phase of fury does not last long. In ten or twenty minutes the worst is over. The froth is replaced by an angry, dark sea with waves several feet high. The white is still there but it is broken and scattered; the waves carry whitecaps but there is space between. The rain is still falling but it is intermittent. It comes in serrate deluges, slackening after each one. During the downpours the water turns silvery from the agitation and in the quieter periods gray again.

Slowly the wind dies, the rain ceases, the waves diminish. In time the clouds break; the sun, then low on the horizon, shines through and the lessening combers are tinged with refulgent gold.

This is the top of the Bay everyone knows, the portion of the fisherman, the yachtsman, and the vacationist. Yet there are aspects of even this familiar scene which escape ordinary attention and which are worthy of considered contemplation.

It is in the hours of calm that the surface is most interesting. Then all the small floating things are visible, then the telltale marks of the Bay's stirring and of the activity above and below are most manifest. The surface becomes a literal mirror; it reflects all the usually invisible surgings and flowings which in more boisterous hours are concealed. It is in times of calm that the long winding tide rips are most evident; if the calm persists, they may often be seen wandering for miles over the smooth surface, plainly marked by the accumulation of small floating things borne there by the water.

One of the most fascinating days I ever experienced was spent exploring the length of one of these rips. The calm that produced it was a protracted one and lasted for nearly twenty-four hours. I was in a small sailboat off the mouth of the Rappahannock River in Virginia, well out in the Bay but still in the region of the maze of pound nets that the Virginia fishermen set in such abundance. It was early spring but a warmth had settled over the water and the still air was pleasantly comfortable. The sun beamed brightly, the atmosphere was clear and sparkling; from the direction of the distant shore came the scent of pine trees, of myrtle bushes, and of spring blossoms; the faint mud smell of the salt marshes was detectible too, and closer was the typical, pleasant odor of the sea, the scent of fish, of hanging half-dried nets, of seaweed, and of all those small indefinable items which mix in unknown proportions to make up the sea smell. In brief, it was a perfect day.

The water had been quiet long enough to become quite clear. All suspended matter had long settled and the Bay lay translucent and bright green. That is, all except a winding yellow ribbon which appeared to the east and curved in a gentle arc to the south where it disappeared in the distance. The yellow band was quite bright and although only three or four feet in width was plainly visible for several miles. My curiosity was stirred and as rapidly as the still air would permit I coaxed the sailboat over to the nearest segment.

When I arrived I was interested to observe that the ribbon was composed of an exceedingly thin film of fine opaque yellow dust. It was very dry and floated lightly on the surface. I collected a quantity on my fingers; it felt waxy, even a little slippery.

Obviously it was no ordinary dust; I was puzzled. The water for yards on each side of the central ribbon was coated with microscopic motes of yellow; they were slowly drifting toward the axis of the curving line. With a knife I carefully gathered a quantity from the very center and stowed it in a vial. Later I examined it under the microscope and its real character was plain. It was pollen, the accumulated life material of thousands of flowers, or perhaps of trees, billions of life-giving cells blown by the wind and there gathered by the silent gliding current in one spot.

I knew that as soon as the slightest ripple broke the slick surface it would all be gone—wasted, drowned, lost—such a prodigious scattering of potential life and wasteful dissipation of it. There, floating, was the effort of innumerable plants, the cause and the reason for the pushing upward of myriad green shoots, the spreading of countless sun-gathering leaves, the opening of millions of petals, the bursting of untold blossoms—all for naught—the pollen so laboriously engendered from air and sunlight and warm brown earth would be destroyed in a foreign element.

It was another of those many instances of natural waste which seem so unaccountable. The phenomenon is legion; every slum has its human debris, dazed and dreary-faced beings in rags and without hope, discarded by the society that spawned them. I have seen butterflies and other insects setting forth in futile migrations, out to sea or into hostile terrain to certain and senseless death. Everyone is familiar with the spectacle of robins and bluebirds freezing in winter weather, huddling in the shelter of trees and bushes, seeking protection from the numbing sleet into which they blundered unknowing on their premature northern journey. The codfish lays a million eggs that two may survive; six of a litter of baby mice are born and die without ever opening their eyes so that the average may be maintained; and men engage in futile wars.

The tide rip was the collecting place of many things besides pollen. Just beneath the surface hosts of jellyfish were jammed together until in places they were packed solid. All the variety of plankton were there, those helpless beings which float without will, aim or direction, which surrender their substance to the waters to be drifted wherever tides and current dictate. So, also, like the pollen were the land things. Mixed with the yellow were numerous little rafts of grainy material, peculiar miniature floating islands, the identity of which for a time escaped me.

It was only when I touched one and it suddenly dissolved, when the individual grains left the surface and went sliding down into the depths and a few stuck grittily to my fingers, that I realized that the miniature islands were really rafts of floating sand grains held only by surface tension and because their individual weights were too small to break the dividing film. How these collections of fragilely poised sand ever got so far away from shore or how they began floating in the first place is not clear. But there they were, many hundreds of them. I suppose they are picked up from the sun-dried beach by tiny ripples and then floated gently away.

There were many feathers in the center of the rip. In

a few moments I picked up, in turn, the down of a duck, the primary of a vulture, some small yellow breast feathers of a warbler of some sort, and the pure white, unmistakable plumage from a gull or tern. The tidal ribbon also contained quantities of insects. There were beetles by the hundreds, spiders and the silk of their webs, innumerable small moths flopping helplessly about, a few butterflies, and most pitiful, a number of honeybees, somehow strayed from their accustomed meadows.

There was no trouble keeping the boat in position. The sails had lost all force and the opposing tides maintained the hull in the exact center. Thus I was able to lie prone on the deck a foot or so above water and observe the life of the tide.

The top of the Bay was a magnet. Jellies, larval fish, crabs, all the undersea beings were crowded as near the surface as possible. It was as though they were fascinated by the light streaming through or were obsessed by an urge to penetrate beyond the screen, to pierce the molten mirror and enter the world of sunlight and air. More logically, the desire to approach the top was caused by hunger, because the surface is the preferred abiding place of hosts of minor creatures, the micro-beings that infest the Bay in legions.

And, in proof of this, a wide school of menhaden swept by in perfect formation, their gaping mouths extended, scooping up the accumulated life. They were skimming the surface as close as possible and they wheeled back and forth following the line of the rip. In the distance I could see ripples caused by the dorsal fins of other feeding schools.

The preference for surface living is marked among many sorts of marine creatures from the smallest to the largest. It is well known that even the deep-sea forms make regular migrations to the top, perhaps for food, perhaps for relief from the ever-crushing pressure, perhaps because some dim instinctive memory recalls the time before the descent of their ancestors into the deeps just as the seasonal migration of birds is believed to be a relict habit begun during the advance and recession of the ice ages.

Warmth may be a factor or the need for light—we do not know. But it is certain that each time the waves are stilled and a calm settles on the water, the sea things swarm to the ceiling of their world. Once, while crossing the Caribbean, during one of its infrequent periods of calm, I saw a vast assemblage of sea cucumbers, holothurians, basking on the surface in the tropical sun. This was an intriguing sight because these beings ordinarily live on the very bottom in the mud and slime. How, in the stillness of the black depths, they knew it was calm and rose to float idly on the surface I cannot say or even guess. The bottom was nearly a mile straight down, yet there they were wriggling contentedly on the gentle swells. And at the first sign of a ripple, when the first wind broke the surface, the hundreds that were everywhere sank silently out of sight and were gone.

I suspect the attachment to the surface may be dictated in part by the very fact that in the sea there is direction and a tangible limit only at the bottom and at the top. All in between is without shape or substance, every inch is like every other. Midwater is but green or blue or black infinity.

Thus life in the sea is a sort of layer-cake arrangement with a limit at either extremity. For most, the tenuous liquid film of the surface is as potent a barrier as if it were the solid polished metal it appears to be from below. And, indeed, that so many of the sea creatures have succeeded in piercing it, to lead the life of amphibians or, like the land crabs, the existence of true terrestial beings, is one of the natural wonders. Only the flying fishes give the surface no heed and go plunking in and out at will.

A number of animals are wholly dependent on the life of the surface and are seldom found far from it except when driven briefly away by storms or other violent causes.

Perhaps the most exact adjustment is that of the skimmers, those odd, black-winged, gull-like birds which are found at the very lower end of the Bay only and up and down the South Atlantic coast. The bills of these creatures have been so altered by their mode of living that the birds are not able to exist apart from the surface of the water. The sight of a skimmer in a meadow is as incongruous as a sloth on a sand bar or a Salvation Army lass in burlesque.

The lower mandible of these strange birds is extended slightly beyond the upper and is designed to act as a narrow scoop. By soaring just above the water, not a quarter of an inch too high or too low, between the narrow channels of the grassy marshes or in the open Bay, or in the lagoons where the water is calm, with the bill scribing a neat furrow, the bird secures its living, literally plows up its food, the crustaceans and small fry that teem at the very surface.

There is exhibited nowhere, by any winged creature, save possibly the hummingbirds, so wonderful a demonstration of aeronautics or of such exact aerial control. For long distances, in sharp or gentle curves, in still air or heavy wind, the long thin sickle-shaped wings hold the body and the bill at the necessary distance. The precision of a skimmer's flight must be seen to be appreciated.

Next to the skimmers, the surface activities of the needlefish are most evident. They are long acicular animals resembling small slender swordfish. They are designed for speed and their torpedo shaped bodies, tapered to either extremity, suggest the engines of destruction they really are. Needlefish in the Bay usually do not exceed fifteen or twenty inches in length and most are much smaller, but whatever their size they are as ferocious of their kind as any denizen of the sea. For the silversides and anchovies and myriad other small minnows that haunt the silver ceiling of the Bay the needlefish are stark terror. Few, indeed, are the victims that escape the arrowlike rushes of these monsters in miniature and the numerous tiny but sharp teeth with which the long double mandibles are armed.

The mode of the needlefish is to lie quiet, as close to the surface as possible, barely moving, or at most idly waving a forked tail. Soon or late some hapless fishlet comes into sight, browsing just under the surface film, feeding on the life still smaller than itself. Suddenly a bolt of silver flashes out of the haze, teeth snap, and the wretched victim is caught crosswise, held fast by cruel ivory. If the victim struggles, the teeth crush tighter; viciously the needlefish shakes its prey, worries it a little, and then when the twitching and the reflex pulling of hurt muscle is over, the jaws are momentarily relaxed. For a second the bruised body floats free in the water; then it is seized head first and quickly swallowed. I have watched needlefish feeding thus until it seemed they would burst; there appears no limit to their gluttony.

Between feedings, the needlefish like to lie motionless,

rocking slightly with the waves, basking in the sun. Their fondness for basking is shared by many other fish. Sharks, particularly, are addicted to the habit and on calm days in the lower Bay it is not at all unusual to see them idling with the dorsal fin protruding just above the water. Crabs, also, like to swim next the surface and their progress can be detected for long distances by the peculiar ripples they make.

The surface of the Bay is a veritable mirror for the activity beneath. The gulls and terns make their living by watching it constantly. They know where the menhaden are feeding, when the rockfish are breaking, where food is to be had for the taking. The birds scan the surface, the wise fisherman watches the birds. When the gulls head for a distant spot, wasting no time in their flight, all the fishing boats do likewise. Screeching, wheeling, diving gulls and agitated water mean feeding fish and a possible good catch. Conversely, a yellowish slick, the oily residue from hundreds of mangled fishes indicates that the feeding is over, that in all probability both the pursuers and the pursued have departed for the depths or have gone to other places.

CHAPTER 13

SAND

AT CAPE HENRY, VIRGINIA, THE Chesapeake Bay and the Atlantic mingle their waters. There, beyond the beach, in a confusion of subsurface ridges and convoluted bars strong currents move in and out, alternately sweeping and scouring the sea bottom. Twice in a day the great flood of brackish water from the inland Bay pours into the bitter salt ocean and twice the ocean tide forces it back whence it came.

Ceaselessly back and forth the opposing currents take up their work of altering the sea floor, grinding and sifting the sand bars, moving on to return again laden with silt and with the residue drifted down from the distant rivers, from the hills and valleys of the remote hinterlands of Maryland and Virginia.

Often great storms deluge the twin lighthouses, beat heavily against the cape, spraying the beach with masses of foam and with the dead and dying bodies of myriad starfishes, sea urchins, broken and dying crabs, the delicate shells of other fragile crustaceans, and the white calcium of innumerable mollusks. With each wave that sweeps upon the beach there is deposited one more creature, one more broken shell, one more ounce of glittering damp sand.

In time, as the never-ceasing waves work back and forth, beating and crushing the residue, this deposited material dissolves into minute fragments, into tiny bits of eroded coarse gravel, specks of lime, of ground-up quartz, silica and resilient chiton. Then, as the storms and currents subside, as the waves gradually diminish, another force comes into play—the powerful all-heating rays of the sun which, evaporating the water, leave the minute fragments to the mercy of still another force, the pervading pressure of the wind. And here begins a whole group of interesting conditions, an ecological progression born of the surging tides, of the grinding surf, and of the heat of a flaming star.

On April 26, 1607, a motley boatload of curious men landed on the beach at Cape Henry only a few hundred feet from the site of the present lighthouse. They were the Jamestown colonists, the founders of the first permanent settlement in North America. There is no exact record of their emotions at this moment but it is likely that their hearts sank at the scene before them.

There was no friendly lighthouse to greet them then, no series of comfortable houses such as now dot the seaside; instead, only a stark shore line of barren sand, of naked dunes which went inland as far as the eye could see, a waste of drifting soil, of scattered dwarfed pines and dried beach grass. And, back of the beach, even as now, towered great mounds of yellow sand, rising into the sky for a full hundred feet or more. This might well have been an American Sahara, a desert on the edge of the ocean, and, as such, it must have impressed the colonists, for in a few hours they continued on their way after first placing a wooden cross as token of their visit.

The area immediately surrounding Cape Henry is one of the most fascinating in the Middle Atlantic states. It is known generally as the Cape Henry desert and is the nearest approach to an actual desert in this part of the country. This vast accumulation of sea-borne, wind-blown sand presents one of the most engaging sights of the Chesapeake region, although not too many travelers and vacationists, in spite of the nearness of Virginia Beach and of Norfolk, seem to be fully aware of its existence.

The "desert" is not conspicuous from the automobile road that passes near the lighthouse reservation but is partially hidden by a scattering of pines and other conifers. To appreciate fully the character of the area one must leave the road and its speeding cars and venture on foot.

It is best when first entering the desert to do so quickly, penetrating immediately to the very heart of the dunes, best to climb to the top of the highest one and there to pause. For then the full magnitude of the place strikes home.

Distant in the background and somber against the gleaming soil is a dark line of green trees, cypress and gum and swamp willow, a line representing only the tops of the trees, for the bases and the upright trunks are hidden by the sand. In places the dunes reach up and up, towering far above the vegetation, completely obscuring the forest beyond. Turning, one finds, in order, more great piles of yellow sand unmarred by any living thing, and still more piles, mound succeeding mound in graceful form until a subtle green steals over the distant dunes marking scattered growths of hudsonia plants and the even more tough beach grasses.

Completing the visual circle, one observes the dunes becoming smaller and smaller, more and more green until they mingle brokenly with a band of pines, bayberry bushes, and other hardy vegetation. Beyond the strip of green plants lies more sand, the smaller dunes of the beach, and finally the brilliant moving white of the breakers and the bright sea green of the Atlantic Ocean. This is the scene.

But the wonder of the desert, the aspect that is most potent, is not the scene but the mood: the desert, like the open Bay, is subject to subtle change. Particularly is this true of the Cape Henry sands, for the sea and the sky determine and govern its mood and are inseparable from it. The dominant sensation is one of somnolence, an air of brooding desolation, yet withal an atmosphere of peace and quiet.

One might say this is the summer mood, for in the winter the northeast gales come charging down upon the Cape, turn the sea and sky into areas of dark gray and somber green. Then the dunes come to life, awaken from their summer lassitude. With the beating of the wind they begin to move, at first imperceptibly, but as the pressure increases in great clouds of smoky sand they inch forward, changing hourly in shape and position, until the very air is full of stinging particles. This is the grandest mood of the dunes.

It is, however, in the more subtle things that the feel of this seaside desert becomes evident. Particularly is this true of that time of year when a drowsy warmth, a promise of the great heat of midsummer, steals over the barren slopes and augments that tiredness which comes from dragging through ankle-deep sand. Then, if one lies down on a big sand pile and drifts into lazy semiconsciousness, there comes an instinctive awareness of some intangible movement which is difficult at first to define.

There comes to the listening and inner ear a gentle singing, so soft as to be only vaguely manifest. There are muted whisperings which seem to run up and down the dunes, rising and falling, as of hundreds of tiny voices. And behind this faint noise will come a muffled roar and sigh, and yet again. Steadily in measured cadence, a throbbing as of some distant and beating heart.

Subconsciously one thinks of great green pines and the sound that comes from the needles when the wind is high, or of the rustle of meadow grass, or of wheat rippling in the wind, of stem against stem, blade against blade.

Opening the eyes one sees that the gentle whispering is nothing more than the drifting sand, of billions of sand grains, blowing up and down the dunes on the east wind, grains rolling and bumping into one another, myriads of infinitesimal sounds accumulating and blending into a soft and continuous noise, very like a genuine whisper. Then, if one looks still farther, beyond the twin lighthouses, there is the surf rolling in and up the slope of the beach, roar and sigh, roar and sigh again, a watery metronome to the musical diminuendo of the sand grains.

To understand the dune country it is best to begin at

the point where the dunes are born. That is to say, at that point where the farthest waves reach, at that line where the longest ripples leave little windrows of dead things, little piles of seaweed and streaks of damp sand.

There the sun has already started its work. The many starfishes cast up by the waves are drying out and shriveling up—those which are not devoured by the sand crabs. The seaweeds fast become dry strings which soon powder into fine dust to be picked up by the sea breeze and carried inland. But most of the sand is coarse and is mingled with bits of broken shell, the carapaces of tiny shrimp, beetle elytra, and all those things which have met their end at the water's edge.

This coarse sand—how odd. Back in the dunes it was fine, almost like powder, made so by the millions of times each individual grain bumped into other grains on its way to the farthest dunes. With each bump a little bit is worn off until the grains become smaller and smaller and more and more round. That such is reasonable is evidenced by empty "pop" bottles left on the beach. In a few days their glossy glassy sides are abraded into opaque material, even the raised letters are rounded and smoothed until they are barely visible. This was the first sandblasting.

From the water's edge the beach slopes gently upward for a hundred feet or more and then rises abruptly in a short dune, the real beginning of the land, for the beach is not true land but is half owned by the sea, which claims it from time to time. This abrupt dune, the continental rampart, is always characterized by a distinctive feature—its seaward face is steep and scoured in graven lines by the wind. Not so the reverse, the landward side.

This slopes gently downward, blending imperceptibly with the land to the rear.

Here is an interesting phenomenon. All the other dunes, with few exceptions, are constructed contrariwise, sloping easily to a crest and then dropping in a sheer fall on the lee side. The sand of these last, on the windward side, is hard and firm to the foot as if packed by the pressure of the wind. But on the steep slope it is soft and yielding, letting one down to the ankles.

Why is this? Because the rampart dunes are eaten away by the winds and by the waves, which cut out the seaward dune faces. The inner dunes, in turn, receive the wind-blown residue of the seaside dunes; this drifts gently up an easy slope until a crest is reached. Beyond, the air is quiet, protected by the body of the dune. There, the blowing sand, relieved of its wind pressure, falls quickly and softly, forming a steep and loosely packed slope.

Just back of the sea-edge dunes the first vegetation begins to take hold, the resistant beach grass. Here in greater or lesser clumps it struggles bravely against overwhelming odds. Out through the shifting sands it sends myriad root stocks, tangles of intertwining fibers which pierce the soil, out and out, until a favorable point is reached where there is then thrust up a hard needlelike shoot which if not too deeply drifted over soon breaks out into a flourishing stem. And nature, as if not satisfied with this ingenious means of ensuring survival, further provides the plant with seeds which in their turn keep the race alive even though the attenuated root stocks may be destroyed and blown away.

It is a graceful sight to see these grass clumps festooning a pure white dune, the long leaves delicately curling over until their tips touch the ground. And, as the tips sway they scribe a circle in the sand, a perfect ring about each plant.

It is to these clumps of grass that many of the hundred-foot dunes owe their birth. The sand drifting on the breeze collects at the base of a clump, piles gently up and falls behind in a little steep drop. Then, unless the wind changes, more and more collects, piling ever higher, over the grass and still higher until a great dune is on the way to completion.

Behind the first grasses other plants take hold. Sandburs, those irritating plants with needle-spined seeds, the hook-clawed cockleburs—and there is a marvel. The cockleburs have a double blossom and a double seed which fuses into a single bur covered with spiny prickles. These double-single seed pods are so timed by nature that one of the two seeds will mature this year, the other months later. Consider the thoughtfulness of this provision. Here in a region of difficult conditions, where a winter's storm may in a few moments ruin all chances of a seed's survival, nothing is lost, for the second half of the waiting cocklebur will be ready a year hence when possibly conditions will be more suitable.

Bayberry plants come next, highly scented, with little gray seeds arranged in clusters. It was from these that the early colonists made their candles, the wax derived by boiling the berries and skimming the surface of the resultant liquid. Many a pioneer tidewater log cabin depended on bayberry for its only evening light; the scent of these primitive candles must have filled the interiors with a pleasant and pungent fragrance. Beyond the bayberry is, finally, a straggling of long-needled pines and cedar trees.

It is the cedars that give the dune country its air of desolation. They are so bent and twisted, all leaning away from the ocean, scarred by the midwinter gales, pointing gnarled branches toward the inland dunes, seeming to cower from the wind that is to come. But really they are brave and they endure and survive where more stately trees would perish.

The strip of cedar and pine is not dense but is broken by areas of open sand and beach grass. It is through these areas that the beach sand pours to accumulate on the farther dunes. Once beyond the trees there is nothing to stop it and there it piles in exquisite forms, on and on, ever higher, in great drifts where even the beach grass cannot take hold. And on it sweeps until the very last and highest dune tops the trees of the untouched forest beyond, a forest of gums and maples, cypress and other swamp dwellers alien to sandy soil.

Here is a sad thing. The great wall of drifting sand inches forward, over the roots, on up the trunks, gently, ever so gently, burying the lower branches, softly encasing the leaves, until one by one they are snuffed out and buried.

In a year or a month or a century the dunes will move on again and will leave behind the dead trees standing exactly as they were overwhelmed, in the precise position in which the sand reached them. There are several of these buried forests beginning to show not far from the lighthouse. They add to the feeling of desolation.

One of the remarkable features of animate life is its determined tenacity in the face of seemingly insuperable difficulties. One would think that in such an area there would be a few living beings. But that is not the case. In spite of the intense heat and the glare of summer, the icy wintry blasts, and the ever-shifting sands and unnourishing soil, in spite of the unstable conditions, a whole host of creatures make the dunes their home and some are specially designed to cope with this precarious existence.

Whenever the weather is warm enough little gray tiger beetles drift about in swarms, becoming more active as the temperature soars. At first the preference of these insects for soft sandy places is not apparent and the answer is not evident until study is given to their life history, to the development of their young. For, when these beetles are in their larval stage, when they are mere wormlike grubs with big jaws, they make good use of the shifting soft sand, use it as a trap to secure their food. This trick they share with another insect of a quite different order, the so-called ant lion.

These creatures excavate little pits by quick motions of the thorax, which tosses the loose grains some distance away. In time, as they work, a little circular depression with steep, loosely packed sides is formed. At the bottom, buried out of sight, lies the waiting larva, jaws open. These larvae are the very souls of patience, for they will remain thus for hours, resting motionless until some venturesome insect—usually an ant, for ants are enormously curious—peers over the edge and mistakenly crawls inside. Instantly the loose walls give way and precipitate the luckless creature to the bottom and into the waiting jaws. And once these jaws become fast they never let go until the pray is dead. Then the larvae greedily suck out the body juices and with a quick flick of the head toss out the empty shell and patiently settle down again to wait for another victim. It is interesting that two insects of quite separate orders with different anatomies should put a single environmental condition to the same use.

Some of the insects that inhabit this barren country seem oddly out of place. This is true of the dragonflies. They are very common, yet all their earlier lives are spent in water, in quiet ponds, and it is to water they go to lay their eggs. Their attraction is the myriad sand flies, little things like gnats but which seemingly carry red-hot needles in their bodies so vicious are their bites. These sand flies are the curse of the dunes and are the only unpleasant feature of the sands. Other unlikely beings are numbers of whirligig beetles and water boatmen, those crazy little creatures which mill around in swarms on shaded ponds and rippling brooks. Finding these in the midst of acres of dry sand is akin to seeing a camel swimming the English Channel—yet on nearly every trip to the dunes I have found their bodies, both alive and dead, lying about in numbers. What brings them to the dunes is one of the mysteries.

Reptiles are common in the sands. Lizards, particularly *Cnemidophorus,* the six-lined race runners, dash from one sheltering clump of beach grass to another or poke more leisurely into the stem bases for luscious insects. Race runners—or sand runners, as they are commonly called—are attractive little beings of dark Vandyke brown bearing six vivid yellow stripes down their backs. They have slender tapered tails longer than their bodies and possess breasts and stomachs of pale ultramarine; their scales are exceedingly fine and their flesh has the texture of soft velvet. They are capable of moving at extreme speeds and, when alarmed, their dashes for safety are so fast the eye can scarcely follow.

Serpents live here in abundance, as is evidenced by the

numerous, queer, twisted tracks they make, tracks stretching out and out like undulating ropes. Except for their tracks they are not, for the most part, too much in evidence. I recall one exception, however, when for a brief period of two or three days the region about the twisted cedar trees seemed alive with the forms of the so-termed hog-nosed viper, *Heterodon.* It was early in the year and I believe they were breeding but cannot prove the point. Nearly every shaded spot contained a viper or evidence in tracks that one had been there a short time before.

This sand-loving serpent is one of the classic examples in nature of the use of sheer bluff to avoid trouble or to intimidate a potential enemy. There are few serpents as mean-looking—and few as amiable of disposition. Beside a hog-nosed viper a copperhead or a rattler appears positively angelic, yet in the world of serpents this snake is the perpetual Casper Milquetoast. I doubt if one could be induced to bite, no matter how roughly handled, and if such an eventuality did occur the bite would be quite harmless.

Hog-nosed vipers seem to realize their inadequacy and, like many inadequate people, resort to bluster and blowing to bolster whatever serves a serpent for a failing spirit. In a way they are in the position of the too-fat man who is in no condition to fight and who necessarily falls back upon diplomacy; unlike the fat man, however, they go to the opposite extreme and pretend they are what they are not. The simile is not inaccurate because their bodies are thick and large and they are unable to progress over the sandy areas they prefer with any show of speed.

Threatened, and with escape a long way off, or apparently impractical, the animal undergoes a Jekyll and

Hyde transformation and assumes its most terrifying aspect. Taking a deep breath it swells the already fat body as much as possible and flattens the head and neck somewhat in the fashion of a cobra. The breath is then expended in a series of angry hisses. This maneuver completes its ferocious appearance and makes a picture of deadly venom.

But it doesn't mean a thing and all the time the poor creature is wishing with all its agitated heart that it was somewhere else. Its only hope is to frighten its enemy so that it may crawl to a safe hole and hide.

But if all this feigned hostility has no effect, the poor reptile still has one trick up whatever serves a snake for a sleeve. A sudden and dire illness seems to engulf the animal. It opens its mouth as if gasping for air; all strength seems to go out of its body; convulsions seize the flesh and cause the muscles to quiver and tremble. Then as these progress it turns over on its back, goes limp, and to all appearance becomes dead.

There is nothing deader than a hog-nosed viper playing possum. Indeed, so complete is its acting that it seems almost overdone, like a love scene in an old-fashioned melodrama. One can pick up the completely limp carcase and carry it around by the tail without the slightest sign of internal movement; the body may even be draped over a limb and it will hang loosely on either side, swaying slightly in the wind. But let it alone for a while, retreat behind some hiding place, and in a little while the head will lift, gaze cautiously about, making certain all is safe. Once satisfied, the creature comes suddenly to life and slithers indecorously for the nearest shelter.

Amphibians are common. Dozens of toads sleep during the day in the grasses away from the heat and venture forth only in the evening when a cool breeze tempers the heat of the sun. One would think they would shrivel from dryness, but they seem to absorb enough moisture from the morning dew and from the bodies of the insects they eat to maintain themselves comfortably. Like the dragonflies, their early lives are spent in ponds and streams—one can see how far they have hopped from their natal watering places—and in the spring when their mating time comes they hop all the way back over the soft dunes to the distant gum forest. It must be a tiresome journey but possibly has compensations in the fat bugs captured on the way.

Just as a new snowfall suddenly reveals the presence of innumerable animals not ordinarily evident, so the Cape Henry sands recount the doings of innumerable dune creatures, particularly those of the night which utilize the dark hours for their activity. Most of the larger tracks come out of the deep woods back of the dunes, show as deep impressions in the yielding slope, but become firmer and more distinct at the top, and then lead off into the dune country. The sand reveals the small humanlike hands of the raccoons, the smaller tracks of the opossums, and even the prints of muskrats, characterized by the dragging tail. What are these last doing away from their backwoods waterways and tidal marshes? About fifteen years ago I found and photographed the tracks of a bear and followed them for nearly half a mile before they turned into the woods again.

And here one can see, perfectly delineated, where some large bird has lit with great force, leaving the imprint of its wings and fanned-out tail; only the talons left no impression, because they closed on the body of a mouse—it was a mouse because its tracks led to the spot and there ended. Just beyond were the four-toed prints of a crow seeking some tidbit in the sand, possibly a fragment of the same mouse. But these signs must be seen before the morning wind comes up, for then grain by grain the tracks will fade away, become mere depressions and dissolve altogether. In their place will be only the marks of the wind, undulating ripple lines like those on the shore when the tide has gone out.

The spirit of the sand country, however, is best shown by the ghost crabs. Colored like the sand itself, they are so much in keeping with their surroundings that they are inseparable from it; their lives are adapted for sand living; the sand is at once their place of sustenance, their home and shelter, their sole world and medium. Ghost crabs apart from white sandy beaches and graceful dunes are as unthinkable as fur-clad Eskimos in equatorial Africa. So perfectly do they blend into their surroundings that they are at times well-nigh invisible. It is their shadowy quality that has given them their common name. They are most fond of the beaches and the dunes close to the sea and there they occur in legions.

I recall sitting one afternoon close to the breakers, so near that the largest waves almost touched my feet. As I watched, the wet soil near my toes began to quiver and then gently broke open. A set of claws appeared, then a pair of glistening dark eyes on stalks. The eyes waved about for a moment or two, saw that nothing moved except the breakers. A second later, like Venus rising from the sea, there emerged from the still-damp soil a ghost crab. But most marvelous of all, the creature was perfectly dry and not a grain of wet sand clung to its entire body. For a time it busied itself opening the hole and carrying out little pellets of wet earth which it deposited radially from the entrance. During all this operation it remained immaculately clean and free of sand grains.

Think of the complications of a life never away from sand either stickily damp or dusty dry. Imagine spending your leisure hours asleep beneath layers of water-soaked grains, of the problems of building a shelter in ever-yielding soil with the walls of your house always ready to crumble away and cave in; imagine your world to be a place where the very ground perennially shifts and changes, blown away by the wind or washed by the surf; think of the difficulty of finding food in this seemingly barren wilderness, and of having a pair of blunt horny claws as your only tools. Imagine, further, that you must withstand blistering heat, the recurring chill of cold salty water, that you must be prepared to survive violent gales, crashing surf, and a host of never-too-distant enemies. Yet this is the normal existence of a ghost crab.

Ghost crabs seem to lead a life of pleasant abandon, scurrying blithely back and forth on their inexplicable errands, dashing hither and yon or creeping like pale shadows over the evening beaches. For their favorite hour is the time of dusk; then they come forth in countless hundreds to prowl over the sands, back and forth between the sea and the dunes.

Their nocturnal wanderings may consume some hours but generally, sometime before dawn, though not always, they will retire to the mouths of their burrows or their seaside dens, their tasks accomplished, to wait for the incoming tide. Then, just before the waves touch the edges they drop inside. Soon a wave larger than the rest rolls up the beach and fills the burrow with sand and bitter salt water. There they rest, buried alive, in the dark and the damp until some unknown crustacean time signal tells them that their hour is at hand again, that all is well in the upper world, and that their lovely beaches are waiting as they always have, for their use and delectation.

CHAPTER 14

BIRDS OF THE BAY

ON A GRAY NOVEMBER DAY I saw on Eastern Bay in Maryland what seemed a breath out of the fabled past. There was a time within the memory of men when the skies over the Bay were literally darkened by the passage of enormous flocks of migrating birds, when across the firmament long streamers of ducks, geese, and swans stretched for miles in seemingly endless lines, passing at times for hours, when on the feeding grounds the massed bodies of fowl obscured the water for acres. The waterfowl still come but they come in diminishing numbers; the sky-darkening flocks are a memory only and the living skeins of birds that once tapestried the sky have degenerated into short, easily counted flights. The years of constant slaughter have taken their toll and the encroachments of a befouling civilization have muddied the waters or poisoned the marshes and the flats until only a residue of the once-magnificent flocks remain to hint of what must have been.

But on this particular morning, in a gray dawn, I caught a glimpse of the glories of the days that have gone forever. All during the previous day a strong cold wind out of the northeast had been blowing, driving low clouds before it with occasional flurries of snow mixed with cold rain. But in the dark, and toward morning, the wind had shifted, moving to the southwest and bringing with it warm air and a dense penetrating fog.

And when the light came the fog hung low over the Bay hiding the buoys and the distant shore line and the mouths of the rivers. By then the wind had died and only the gentlest of ripples lapped against the sandy beaches. Because of the calm and the clinging damp of the fog it was very quiet, all except a faint murmuring which seemed to come out of the shrouded distance. At first I was not sure I heard it, but then it came again, clearer but still muted and indefinable.

My curiosity was stirred and I moved in its direction, toward the pine-clad tip of Kent Island, toward Bloody Point and the shallows that extend for long distances from that point and break the sea coming in from the open Chesapeake. Presently the sound became more distinct, an odd gabbling and a queer rustling and splashing. The noise appeared to cover a large arc of the invisible horizon and seemed to emanate from beyond the beach, out in the open water.

When I came close, when the individual sounds that made up the whole began to become distinguishable, the fog began to lift a little and there appeared not far from shore a white line as though foam was piled deep in billowy drifts. The line was so white against the gray that it seemed to give off a light of its own, a loom much like the reflection of snow. Then the fog lifted completely, and against a dark-green sea and a gray sky appeared one of the largest flocks of wild swan I have even seen. There must have been nearly eight or nine hundred birds and they were massed together between the deep water and the beach. They were still a long way off but through my binoculars I could see individuals preening themselves or feeding. Others seemed just to be idly drifting about. The sounds I had heard were the combined splashing of hundreds of bodies as the birds raised and lowered their heads to feed, the accumulated whispering of countless feathers being rubbed together and the low notes of their voices. These last possessed a vibrant quality much like reed instruments in an orchestra; the low deep tones of the clarinet most nearly approach it.

Presumably, they had all come in during the storm and then had moved toward the quiet waters in the lee of Kent Island when the wind shifted. I had the feeling that they had just arrived, for they seemed unduly noisy, "talking" exuberantly as swan are known to do after completing their long journey from the north. Down the crests of the Appalachians they had come, from mountaintop to mountaintop, winging along at nearly a hundred miles an hour, swooping high above the trees with the wind blowing strong behind them. And then when they had reached the broad ribbon of the Susquehanna they had probably veered east and poured down the valley, over the dammed-up river at Safe Harbor and Conowingo, over the lights of Havre de Grace far below, and down through the night along the wind-tossed upper Bay to the quiet waters off Bloody Point. And it is likely that there the first faint gleam of daylight caught them and they had glided in, flock after flock, to rest and feed, hidden and kept safe by the enveloping mist.

Captivated by the unexpected display, for I never expected to see such a sight again, I circled inland over wet fields and through fog-dampened woods to approach the flock through a screen of dark-green pines and holly. By creeping through sheltering vines, from trunk to trunk, I managed to arrive undetected at the top of a small bluff overlooking the center of the gathering.

As I reached the edge and sank into the vegetation out of sight, a wonderful event occurred. At that very moment a rift must have opened somewhere in the clouds, for a long shaft of soft rose-colored light stole down through the overhanging mist and bathed the whole group in a luminescent glow. Against the somber background of gray sky and dull-green sea the sight of these hundreds of graceful, clean, curving bodies suddenly lighted with brilliant pink was exceedingly beautiful, and I rate it as one of the most enthralling spectacles I have ever witnessed. For a little while the matchless scene endured and then the color faded and the world was once again a monotone.

Such sights are so rare today that they may happen only once or twice in a lifetime, yet they were once commonplace. However, even now one can still find in the Chesapeake region, in the proper season, as many swan as in any other place in the United States. Flocks of one or two hundred birds are frequently observed in the rivers and estuaries of the upper Bay and they are relatively common all through the winter months wherever there is suitable vegetation on the tidal flats to attract them. They occur in the greatest numbers on the Eastern Shore in the vicinity of Eastern Bay, Chester River, and in the reaches of the Big and Little Choptank estuaries.

The Bay is the gathering place of a great number of other birds and there are few places in the eastern United States where one can see, either regularly, occasionally or periodically, as many varieties as in the Chesapeake region. The Bay is an ornithologists' Happy Hunting Ground. Because of its geographical location midway between the North and the South and because of the diverse character of its innumerable creeks, rivers, and estuaries with their great variety of bordering vegetation, suitable conditions prevail for the support and maintenance of a large assortment of birds of assorted types.

At the Bay's mouth true oceanic conditions prevail; there are wide expanses of open salty water offering abundant food for such marine birds as gulls, terns, cormorants, gannets, and petrels. Elsewhere there are thousands of small winding channels, innumerable quiet rivers and narrow creeks, lagoons and ponds of all sizes, shapes and salinity offering haven for ducks and grebes, for loons, rails, bittern, and herons. Miles upon miles of fresh and salt marshes support a varied population of swamp birds—red-winged blackbirds and marsh wrens. Long stretches of sandy beaches attract legions of sandpipers and other shore birds; cliffs and sandy bluffs are the homes of kingfishers and bank swallows. And bordering these hundreds of miles of river, marsh, and open Bay are large areas of forest land, of mixed oak, maple, poplar, and other deciduous forms in the west; pine, cedar, myrtle, holly, and other evergreens toward the east. There is open dry terrain of waving grass, and also acres of soft yielding soil replete with ferns, mosses, and lush vegetation. All these have their special bird populations, and almost every important group of North American bird is represented in abundance by one species or another.

The Bay is a center also for another of our magnificent American birds, the Bald Eagle. With the exception of some parts of Florida there are probably as many Bald Eagles in the Chesapeake region as any other area of comparable size in the eastern United States. The Bay offers them food in plenty and enough wild country to live relatively unmolested. True, each year they are being crowded a little more and their living room is being diminished but there is still a goodly number remaining. Some have even managed to persist in spite of the encroachments of summer cottages and the accompanying disturbances of resort life. Until a couple of years ago a thriving family of Bald Eagles maintained themselves on the south shore of the Magothy River in Maryland in the midst of a noisy colony of houses only a few hundred yards from their long-established nesting tree. Strangely, so preoccupied were the majority of the colonists with their special pleasures that only one or two families were aware the eagles were around.

However, some benighted character shot one of the pair and left the body beneath the tree to rot. This disaster was followed in a few months by the nest falling out of the tree where it had rested for innumerable seasons. But not long after, the surviving bird was back with a new mate and the eyrie was rebuilt nearby on more secure foundations. As far as I know it is still in use.

On the Western Shore between Baltimore City and the Patuxent River I have personally known of some twenty eagle nests and it is possible there are several times that number of which I am not aware. For obvious reasons I prefer not to divulge their exact locations; there are too many trigger-happy individuals who would be only too glad to add an eagle to their list of trophies even though the risk of a severe fine is considerable.

However, there is one center of which I will speak in the hope that by some remote chance some further thought will be given by the state of Maryland to preserve the particular area in its present condition. The district has already been considered as a state park but at the time of this writing the project appears to have been abandoned and the region is being rapidly turned into another of the ubiquitous shore resorts.

In Maryland, between Cove Point on the Chesapeake and Drum Point in the mouth of the Patuxent River, is a large area of heavily forested land of splendid mixed oaks, gums, and pines. Ecologically, the region offers a unique variety of vegetation and special conditions, and it harbors a diverse collection of wildlife. There are several well-established heronries, a large collection of muskrat houses, and a flourishing tern nesting site. But most important, the deep woods shelter the eyries of a surprising number of eagles. Just a few years ago, I remember sailing slowly along the water's edge admiring the beautiful curving sand beach that borders the wild rice and cattail swamp that lies just north of the Cove Point lighthouse. Glancing toward the sky I perceived a number of very fine dots, so high above the trees as to be almost out of sight. They were hanging motionless in the air and at first I thought they were vultures. For a long time they remained, circling slowly far up in the blue, and then for some reason unknown to me, the outermost dot began sliding earthward, at first slowly and then with increasing speed. It was followed at intervals by the others, and as they rushed down I perceived they were not vultures at all but full-grown eagles. Without a halt they continued more and more rapidly. Then just when it seemed they were about to dash into the earth they checked their flight, curved gracefully into the air again, wheeled, and then settled in some pine trees overlooking the swamp. There in one group were more eagles than the average person sees in a lifetime.

The sunlight caught their white head and tail feathers and I identified six of the birds as adults; three others were brown all over and were immature birds, probably less than a year or two old. So many eagles at once seemed most unusual and I had not previously thought of them as moving in groups. In later times I saw them frequently in this part of the Bay and congregations of four or five were not uncommon. However, as I observed later, most of their hunting was done singly or in pairs. When the latter occurred they acted as efficient teams, particularly when trying to steal some poor fish hawk's newly caught dinner. In such cases the fish hawk was harassed continually, first by one bird, then by the other, until driven to desperation it dropped its catch for the sake of peace.

The Chesapeake Bay eagles feed for the most part on fish either stolen from the ospreys or caught directly by diving. They dive from considerable heights and hit the water with a resounding splash. While they are quite capable of securing their own prey, they seem to prefer to take it from the fish hawks, this being much less trouble than plunging into the water after it. Almost every fish hawk colony will have its neighboring eagle waiting for tribute. The eagles are in the position of the robber barons, lordly in bearing and mien, true-blooded aristocrats but thieves at heart.

The Bald Eagle is our national emblem and it may be a little disillusioning to note that besides being a robber the symbol of our national dignity is also a carrion eater. It is not at all unusual to observe eagles on the Chesapeake beaches devouring the dead carcases of cast-up fish. This is a reflection, I believe, of their essentially lazy nature; only rarely have I seen eagles indulging in wasteful effort as so many other birds do. They spend a great deal of their time perched on branches of tall trees idly viewing the landscape or just dozing. The rest of their waking hours are utilized in soaring, often at incredible altitudes.

However, in full action an eagle is a never-to-be-forgotten sight. I remember a fall day on the Potomac several years ago. It was one of those times when innumerable flocks of ducks were moving about, skimming in groups a few feet above the water. One such flight, of about thirty individuals, was moving downriver when suddenly every duck, as one bird, veered and then plummeted to the surface of the water. They landed with a simultaneous splash and then scattered in all directions. And not a moment too soon, for out of the seemingly empty sky came hurtling the form of a full-grown eagle. The big bird was plunging from the heights like an arrow, white head and tail flashing in the sun. It had been aiming at the center of the group and the sudden dispersal of its selected prey temporarily disconcerted it, for it quickly veered skyward again, mounted to a hundred feet or so, and then with broad beating wings took off after one of the fowl which by this time had burst into flight again and was fleeing for safety as fast as it could go. But the duck did not have a chance; with ease the eagle overtook its quarry and then once again plunged. The duck and the eagle hit the water together—a mass of flying spray and vibrating feathers. With a tremendous surge the eagle cleared the water and with the crumpled body of its victim went flapping toward the Virginia shore.

I believe, however, that the Chesapeake eagles seldom indulge in such activity if there are fish available and that the number of ducks or other game fowl taken is relatively small. About their nests I have found the skulls and bones of a few rabbits, a muskrat or two, and in their pellets, in which they cast up the indigestible portions of their meals, the fur and residue of squirrel carcases. Nevertheless, one nest had beneath it part of the pelt of a grizzled old raccoon and, knowing the fighting abilities of this mammal, I can imagine the eagle must have had its talons full when it made the capture.

Most of the nests I have examined have been in living trees, principally good-sized oaks or pines. Generally they are located from a quarter to a half mile inland in deep woods, often on the edge of swamps. The nest itself, depending on its age and the activity of its owners, can be an enormous structure. Years ago there was a nesting tree on Furnace Creek near Baltimore which a friend of mine climbed seeking eggs for his collection. It was situated at the very peak of a tall oak and when he had scrambled to the top and over the rim his body was quite invisible from below. He told me he was able to stretch out full length with ease. He should know because he spent the better part of a day in the nest. After climbing over the rim, which was nearly a hundred feet above-ground, he could not see below again to secure a foothold. The curve of the nest hid the only branch capable of holding his weight. After lowering his body several times in an attempt to find it with his foot and almost falling because of the decayed nature of the branches on top, he lost his nerve and became too frightened to attempt a further descent. He stayed in the tree for a whole afternoon while someone went home and found a rope to rescue him. This tree is long gone and its site is now surrounded by chemical factories.

Some eagle nests, like those of ospreys, are in use for many years and are added to each season. There is one near the mouth of the Potomac which I know has been inhabited in season for over twenty years, and there is at least one record of an eagle nest in use for over thirty years. The Magothy River nest was at least fifteen years old before it fell.

The nests are constructed of sticks, some of good size, and lined with grasses, seaweed, and sometimes mosses with a scattering of feathers on top. Normally two dull-white eggs are laid, sometimes three. The incubation period is a little over a month and both sexes share in the hatching and care of the young. It is reliably reported that if the birds must leave the nest for any period of time, the eggs are carefully covered with grasses or feathers so as to conceal them until the parents return. Ten or eleven weeks are required for the young to acquire their immature plumage and be ready for their first practice flights. Nearly four years elapse before they are fully adult and possess the distinctive and striking white heads and tails.

For all their large size and reputed fierceness, the

Chesapeake eagles are often harassed by birds much smaller than themselves. Crows love to bedevil them, as do other large hawks, and several times I have seen individuals from the Cove Point nests flying majestically along surrounded by a horde of screaming diving crows. But they are dignity itself and they rarely appear to notice their tormentors. However, I once saw an eagle so annoyed and worried by a kingbird that it quite lost its normal poise. Kingbirds are tiny fellows and are about as harmless as birds can be, but for some inexplicable reason they are imbued with an insane desire to do battle with any other bird that comes along, and the bigger the better. They take an unholy joy in attacking hawks, owls, or trespassers on what they consider their particular domain. Their energy is unbelievable and their lack of caution inexcusable; yet I have never seen a kingbird avoid a chance to undertake the humiliation of some creature much its superior. Maybe they have an insatiable ego to pacify. Most strange, so audacious are their forays and so persistent their irritating, if innocuous, attacks that larger birds often lose all aplomb and flee precipitately.

On this particular occasion, an adult eagle was skimming above the beach north of the Cove Point lighthouse, strictly minding its own business, when out of the trees burst a screeching kingbird. With fury it flew at the eagle's head, striking at its eyes, and flashing around to dive again. It seemed beside itself with anger and in comparison to the massive eagle it looked like a mosquito gone berserk. At first the big bird ignored it, except to flinch a little or to twitch its head out of the way. But this forbearance did it no good, for the kingbird renewed its attacks, whirling, diving, lighting momentarily on the eagle's back only to dart away again and dive some more.

For some time the eagle flew straight ahead but then the continued irritation began to wear on its nerves and the bird tried consciously to shake off its annoyer. It even flew a distance up into the sky in the vain hope that the obsessed little creature would tire and return to its accustomed woods.

Then I witnessed what appeared to be one of the most calculating pieces of planned strategy I have ever seen executed by any animal. While high in the air the eagle shifted into a smooth gliding flight as though it were tired of the whole business and preferred to ignore its difficulties. For quite some distance it maintained its pose and all the while the kingbird continued its perverse activity, becoming increasingly insulting in the closeness of its attack. Then, when the bird became confident to the point of carelessness, without warning, the eagle suddenly turned upside down, rolling over in the air with amazing speed. Through my binoculars I saw a single talon flash out and grasp the annoyer. The chorus of screeching and chirping stopped instantly. The eagle then righted itself and continued its soaring with one foot hanging. For several minutes it held it so and then turned and circled over the water. As it came closer I saw the talon relax and a tiny ball of bedraggled feathers flutter to the Bay beneath. Without a backward glance the eagle flew inland and disappeared behind the trees.

Next to the eagle, the Chesapeake bird that possesses the most natural dignity is the Great Blue Heron. The eagles inspire respect by the very fierceness of their mien, by their aristocratic carriage, and by the lordly expanse of their broad wings. The herons achieve it by quite different means, by the quiet deliberateness of their motions, by their stoic indifference to the world, and last but not least by the strange, slightly weird beauty of their form. A heron's life is an unending succession of graceful poses; they were meant to be painted or sculptured. The frequency of their delineation on Japanese and Chinese silks and tapestries is no accident; it is the result of the natural appreciation of highly artistic peoples.

Of all the Bay's birds I think of them most often. There is no corner of the Chesapeake where they are not found, no quiet swamp which does not have its heron standing motionless on one leg, its flowing curves reflected in the still water. They frequent the pound nets, posing on the stakes, waiting with the patience which only herons know for some fish to come close. It is rare that the summer sky does not reveal the majestic silhouette of a Great Blue Heron against the setting sun moving to its nocturnal hunting grounds. Herons stalk the beaches and the tidal flats or stand like living statues on the fallen, whitened logs, impassive, unmoved by the restless world about them. They are not only a part of the Chesapeake scene but the very symbol of it; they are the spirit of the marshes, of the waving cat-tails, of the winding channels, of the water-lily beds, of the dark quiet pools, the bordering pines, and the winding yellow beaches. No other creature fits so perfectly into the prevalent mood and seems so naturally to fulfill the atmosphere of the Bay country. In the full summer when the dead calms fall upon the waters and the marshes quiver with heat, when the water is glassy smooth and the cat-tails are still and quiet, no sight so perfectly completes the picture as a Great Blue Heron posed in complete immobility against the green reeds. At such times the whole world broods and the herons brood with it. Like certain ancient Oriental gods, they seem to contemplate some abstract infinity, some deep unutterable mystery, or some very old and sorrowful truth. Like the mystic yogis, they seem to achieve a sort of avian Nirvana, a motionless state of detachment, in which all external existence is extinguished, all desire banished, in which complete rest is the total of achievement.

And they are also the spirit of the night, for the night is their feeding time and it is then they break off their midday dreaming and go stealing through the shadows like creeping ghosts. In the wan moonlight they cautiously stalk up and down the marsh channels, moving ever so slowly, warily lifting their feet and setting them down so as not to make the slightest ripple. Then, if the sound of their darkened world is not to their liking or if they feel their presence is detected by their intended prey, they will freeze into position, merge into the gloom, and instantly become absorbed by the night.

Great Blue Herons have an air about them of something ancient; they are truly a relic of the past, not the immediate past like the swan of Bloody Point, but of the years of the geological eras. The feeling of great age, of antiquity is imparted in some intangible manner, by the way they move, by their strange, detached deliberation, and by the very shape of their bodies. They do not quite belong in a modern world; they seem to conjure pictures of antediluvian vegetation, of tropical flora, of hot sun and steaming primeval jungles. I did not fully realize how very ancient they appeared until one night I saw one fly across the face of the moon. The black outline of its body could have been lifted entire out of one of my paleontological textbooks; it was the silhouette of *Pteranodon*, the extinct flying reptile with the twenty-foot wings. But for the indisputable fact that it was the seventh moon of the fortieth year of the twentieth century, the scene might well have been any night of the Cretaceous period forty million years ago.

This is not so fanciful as it seems, for the Great Blue Heron, along with all others of its kin, is a primitive type of bird and in all the countless centuries since the fantastic Age of Reptiles when the first birds appeared on the earth, it has not managed to rid itself of all its reptilian characteristics. There are a number of features of its anatomy which provide unquestionable testimony of its heritage. Most obvious is the evidence furnished by the legs and feet; this is also borne in various degrees by all birds of every type and station. Bird legs are covered with well-developed scales; these differ in no important respect from the scales on a snake's back or on the limbs of a lizard or turtle. All birds, including the Great Blue Heron, may be viewed as modified reptiles become highly specialized to fit special circumstances. This statement, I am sure, will serve to alienate the regard of numerous people otherwise well disposed toward birds; nevertheless, the evidences of anatomy and of paleontology are too potent to disregard.

Indeed, the only major anatomical feature, beyond certain important differences in the circulatory system, which distinguishes the herons and other birds from their reptilian ancestors is the possession of feathers. No other form of life possesses feathers; in this one characteristic the birds stand alone. Yet it is believed that the first feathers were developed from scales, and a study of the gradation of feathers on birds' legs lends credence to this belief. In certain birds, including the herons, every stage of development from scales to feathers is represented.

Even in their reproduction the birds, and particularly the herons, are remarkably reptile-like. Both reptiles and birds produce similar eggs. Those of the Great Blue Heron are singularly primitive; their peculiar elliptical shape is almost exactly duplicated by the eggs of a number of reptiles.

In the Chesapeake Bay region the Great Blue Herons nest singly or in small colonies, generally in deep woods at the ends of marshes. The nests are fairly large affairs of loosely piled twigs and sticks. They are usually rather flat on top and in a high wind one wonders what keeps them in place. Commonly there are four greenish eggs and the incubation period is about twenty-eight days.

The newly hatched herons are horrible-looking travesties of their parents; their behavior is atrocious and they are awkward, ungainly, and quarrelsome. They are so unattractive that it does not seem possible they will mature into the graceful, beautiful birds they inevitably become. While very young they are fed by both parents. Their early diet is soft, regurgitated, partly digested and very, very odorous fish. They are usually fed in rotation but the most aggressive get the lion's share. Their feeding is rather ludicrous. The old bird approaches the nest and stands for a while on the edge with head held high. Then it is lowered with a series of peculiar pumping motions; instantly the bill is seized by the waiting youngster and a wrestling match ensues until the food is either spilled over the nest or flows into the young one's mouth. While all this is going on the other birds are fighting for attention or squabbling for preferred positions.

The Bay herons feed on fish and frogs and they secure their prey by still hunting or stalking. While waiting for fish to come within reach, they will stand for hours, not moving a muscle or twitching so much as a feather; they become graven images. So long are they capable of remaining still that I have never had the patience to wait and determine the extent of their endurance. Time to a heron is nonexistent; they are unconscious of and completely indifferent to its passing. Only the approach of a suitable victim will stir them. Then with lightning rapidity they come out of their torpor, the bill lashes out, and the prey is seized crosswise or impaled by the sharp swordlike beak. After capture it is flipped into the air for a second, caught, and swallowed head first. If the fish is a very large one, the heron may take it to shore and beat it to death on the ground.

Great Blue Herons have no important enemies except man. Like the eagles, they are sometimes worried by smaller birds but there are few animals which would dare to attack so courageous an antagonist. If wounded and at bay, a full-grown heron is highly formidable, its powerful sharp bill can inflict serious wounds. But they are peaceful beings and ask only to be allowed to roam their preferred solitudes and to have freedom to seek their food in their quiet, time-hallowed, patient way.

When the swan and the eagle pass from the face of the earth, as they will before too long if their numbers are continually diminished, it is likely the Great Blue Herons will be still posing in their inimitable and graceful fashion and contributing, as they always have, to the beauty of the Bay country. For all their ancient heritage, they are highly intelligent beings; and within the limits of their particular specialization they have achieved a success which has tided them over several thousands of centuries. As long as there are marshes to shelter them and fish on which to feed they will endure, and because of them the earth will remain a more artistic place, and the Bay will retain a little more of its old-time loveliness.

CHAPTER 15

THE BOTTOM OF THE BAY

LIFE IS ESSENTIALLY HORIZONTAL or lateral. The total of earthly existence may be viewed as a tissue of stratified protoplasmic film thinly and sparsely covering the surface of our planet. It is restricted at most to twelve miles of vertical distance, half of which is a mixture of finely proportioned gases, the remainder a skim of slightly enriched soil and carefully compounded liquids. And in this limited zone it occurs in narrow strips sometimes of astonishing thinness. Even man, the most widely distributed, the most versatile and precocious animal of all, is bound by five miles of altitude;

most of his astounding activities are confined to half this distance and he is like a crab which is able to progress with ease only sidewise or forward or backward.

The Chesapeake Bay is about one hundred eighty miles in length and approximately twenty to thirty feet in average depth. Yet most of its life moves within a few inches of the surface or within three feet, one way or the other, of the bottom. Only the micro-animals and the helpless drifting plankton live throughout and even these often sort themselves into narrow bands. Thus the life of the Bay may be likened to a sort of multiple-tiered birthday cake or an animate sandwich. Most of the larger forms frequent the bottom. They may leave it from time to time for various purposes, but it is to the sea floor they return for their principle activities, for their sleeping and reproduction, their sustenance and daily routine.

Every Bay fisherman knows this and it is no accident that he does not hang his hook in midwater or dangle his line at random. The expert angler adjusts his lure to a few inches above the mud or sand and it is there he will eventually make his catch; the hook that is baited too high is usually left untouched. In an academic sort of way I had long known of this stratification but it was not until I began seriously to explore the depths of the Bay from the fish-eye level that I became fully aware of it.

However, my first real attempt to concentrate on the floor of the lower Chesapeake so unexpectedly startled me and left me so shaken that for a short time I would have abandoned the project gladly. I had submerged in about thirty feet of water near Gwynn's Island, Virginia, and was standing on the edge of a steep sandy depression which dropped mistily away into the somber green depths. By lying almost vertical against the slope I was able to keep my eyes comfortably on a level with the sea floor and well out of reach of the strong tide pouring in out of the Atlantic Ocean only a few miles to the east.

I had been on the bottom for a rather long time and for some ill-defined reason began to feel uneasy. It is a sensation which, in some degree or other, I have always had when undersea, and it is attributable, I think, to the strange character of this unaccustomed element. There is a finality about dropping beneath the quivering, tenuous film of the Bay's surface that has to be experienced to be fully realized. The moment the top is passed, one leaves behind sound and scent, loses three-quarters of one's customary vision and retains only a heightened sense of touch. I think I know now how the blind must feel when they reach that sickening stage just before total blackness closes in, when the world becomes dim, when objects fade slowly into shapelessness, when the wall of imperceptibility presses closer and closer.

For thirty feet the trip beneath the Chesapeake Bay is into semiblindness. The skin quivers when brushed by invisible currents and softly stroked by small animalcules; the slightest changes in movement or shape are carried to the brain accentuated many times. Shadows assume unreasonable proportions and moving forms appear larger than they really are.

I have dropped beneath the surface of the open ocean off the coast of Florida and have explored the fantastic and incredible coral reefs in the West Indies and have there stood on underwater cliffs where the ocean floor dropped down for hundreds of fathoms into the depths of the unknown. I have seen huge fishes in those tropical waters, shark, manta, and barracuda, but never until I became acquainted with the depths of the Chesapeake

Bay only a few miles from my home had I felt so wholly separated and apart from the outer world.

In the tropics the waters are crystal clear, and even at the greatest depths there is a lightness and beautiful quality about them. But in the Chesapeake one is surrounded and enveloped in gloom; this imparts a sense of helplessness and uneasiness which has no basis in fact. The intangible is simply more fearsome than the real.

The feeling is induced, in part, by the abundance of stinging jellies and the instinctive reflex from the sight of their ever-present forms. With the tide they were sweeping by just over my head and their tentacles were sliding uncomfortably close. Then, too, only a few moments before a large nereid had slithered over my bare leg. I had disturbed it with my feet and it had burst from its den of sand and, harmless though it was, its bristling flight and its iridescent reddish segmented vermian form had made me jump. No one likes a six-inch worm complete with tentacles and cirri crawling over one's exposed and sensitive flesh.

And then, as if the worm was not enough, the sand on which I was precariously resting softly dissolved and I began slowly sliding into the depths. Like an actor in a slow-motion cinema, I skidded helplessly down, unable to stop my earthward flight, tearing up clouds of black impenetrable silt as I went. Though I knew the air hose would check my fall, the hopeless feeling engendered by my inability to secure the least purchase on the yielding soil augmented the taut condition of my already overwrought nerves.

So, when I once again regained my position on the channel's rim I was in no condition for further experiences nor prepared for what was to follow. While I was laboring to steady myself against the pressure of the current and trying to cling to the sand, peering vainly through the haze created by my fall, I saw far off a great gray shadow approaching rapidly just off the bottom. I had never seen anything quite like it or as strange in shape. Like a large swooping bat it glided over my head, obscured the light, and then fell in clinging folds over my recoiling body.

For a horrible second I pictured an attack by a monster devilfish, for that was what it seemed to be in the murky water. With my arms entangled, my blood suddenly pounding through my ears, I felt myself hopelessly imprisoned and then falling unchecked again into the depths. All sorts of ghastly thoughts went racing through my mind as I fought to free myself from the clinging embrace.

Then, with a tremendous surge of relief, I breathed again and knew my fearsome assailant for what it really was. It was only the awning from the boat, which somehow had fallen overboard and which, fully extended and carried by the tide, had dropped over my cringing shoulders. But so damaging was the experience to my nerves that it was only by the exercise of a great deal of will and good common sense that I prevailed upon myself to try immediately again. I consider the adventure of the awning my most terrifying underwater experience.

It is this habit of the Chesapeake's underwater residents of suddenly appearing unannounced out of nowhere that is their most unsettling trait. For all the number of descents I have made I still have never become quite used to having unexpected creatures materialize before my very nose without warning. It is a situation never encountered in the tropics; for there, unless one is approached from behind, one's undersea visitors can be seen for some yards and their arrival anticipated.

Happily, the next callers after the awning were more reasonable in shape and their size more in keeping with the state of my sensibilities. In fact, so lovely were their forms that I temporarily forgot my agitation and focused my full attention on their passing. I first saw them as flitting shadows which, moving with the tide, swept past my sandy declivity and then vanished down the slope. The sand was still disturbed by my groveling and apparently the swirling silt bothered them, for they avoided me. But a few minutes later, when the smoky mud had all been carried away, they approached without fear and I saw them for what they were.

They were adult butterfish, and if ever by some unexplained miracle I become endowed with the ability to set down on canvas the more lovely sights of this dark world of underwater, the butterfishes will be first on the list. As I saw them that day they were skimming just an inch or two above the sand, their curved ventral fins just missing the bottom. There were several long files of them and they were moving in slender undulating lines only a few individuals in width, sweeping in graceful arcs just above the sea floor. Many were moving in pairs and their delicate forms maintained an exact duplication of action. If one turned, its companion did likewise; even the fins quivered and bent in unison. These may have been mated pairs but this I do not know, for there was no means to tell short of actual dissection.

Their bodies were the ultimate of grace and their shapes were sheer artistry. Highly compressed vertically, they presented, head on, an exceedingly thin and slender oval; from the side they were a series of beautifully molded arcs and tapered sweeping lines. They were the epitome of streamlining and even when quiet suggested flowing motion. Their shapes represented without flaw that sort of grace which produces, in different form, those elements which yacht designers dream about, the undisturbed curve of entry into water and the easy diminishing wake astern.

But it was not in sculpture but in the realm of color that the butterfishes remain in memory. They were pure, refulgent, shimmering gold. Not the gold of precious metal or even of the sun, but a live, quivering iridescence which has no counterpart in the upper air except rarely on the wings of certain tropical butterflies. It was of a hue and texture which is found exactly duplicated nowhere else on earth, and when set, as it was, in its proper surrounding, the misty world of dark-green water, it was heightened and accentuated by its subdued background.

The sight of these graceful fishes, winding through their half-lighted world, a procession of smoldering jewels, was, like so many things of the undersea, unexpected. I had seen their forms many times in fishing boats and on smelly market stalls but never had they given hint of their surpassing beauty. And as I watched, one group wheeled, caught the light streaming down from above at a slightly different angle, and changed momentarily from gold to flaming lavender. They persisted for some time, became less and less, and then disappeared forever. Where they were bound or why will never be known. Probably, because they were adult and it was early summer, they were going to lay their eggs in whatever special spot in the Bay they had selected. I do not know, and I never saw them again.

The fishes of the sea, like the beasts of the land, have

their roads, their special trails and paths along which they prefer to travel. While the whole sea bottom is open to their movement, there seem to be choice avenues for their comings and goings. These passages are not clearly defined and there are no marks or signs to show the way. I remember once standing on a white sandy bottom in the clear tropical waters of the Bahamas midway between some moss-covered cliffs on shore and a dark, drowned subterranean ravine. For some reason there was a constant traffic between the cliffs and the depths. A wide variety of fish, great and small, were filtering back and forth, singly or in groups. They could have gone in straight lines in either direction, for there was seemingly nothing to bar their way. But almost without exception they worked their way along the cliff wall to a point near a large rock and then took their departure for the ravine, or did the reverse coming back. The path, oddly, was not straight but went in a wide curving arc. I moved over to the place hoping for some hint or reason but found nothing. These underwater roads must be defined by other marks than obvious physical signs; their borders may be determined by small changes in temperature, salinity, or oxygen content. Smell or vibration or some other unknown quantity may fix their limits. We do not know.

But that these highways exist is certain and, once found, they repay close attention. My location off Gwynn's Island must have been in the very center of one of these, for at that one spot I observed more migrating life than in all the other places I visited together. I marked it with a small buoy so that I could find it again and in the two weeks during which I made it my favorite underwater haunt it seldom failed. It was on the very edge of a thirty-foot contour line, one which extended with only a few breaks almost to the open sea. The contour was sharply defined and I made good use of its sudden drop to deeper water to view the sea bottom at a comfortable level. This is not always easy to achieve, for on the open floor one is hard put to remain still and is in constant danger of being swept helplessly away by the tide.

The contour fell abruptly away to about forty-five feet to a bottom of soft clinging ooze. Though I examined this lower area carefully, it was always singularly devoid of evident life, not counting the ever-present diaphanous ctenophores. The same was partially true of the gentle sand slope reaching from the contour rim to the distant beach. But the very edge, except for certain hours of the day, was almost always peopled with marine creatures going or coming. And like the butterfish, without exception, they hung close to the sand, gliding only a foot or so at most above its well-swept surface.

The marine travelers fell into several castes or categories, and in their traveling habits were extraordinarily like people. There were some, and the butterfish were of this type, which invariably journey in lines or queues. They moved along tail to tail, or in following groups of twos and threes, seldom far apart. Some of their lines extended for dozens of feet and some moved in parallel rows. I have seen the same on city streets and they reminded me of the straggling lines that automatically form when the big office buildings let out, of the attenuated but definite group that sets out for the nearest corner bar and the equally impatient serpent of individual motes that is heading for the subway or for the Long Island trains.

Then there were the lone hurriers, those impatient individuals intent on their solitary journeys, which burst out of the green haze and into the mist on the other side. They always seemed to have the very devil on their tails, for they appeared and vanished without pause, coming and going so rapidly as to leave one wondering if they had ever appeared at all. Such was the brief appearance of the only black bonito or crabeater I ever saw. Although I had only a momentary glimpse of its rakish streamlined torso, there was no mistaking it; the long dark stripe running the length of its body identified it without question. As its name suggests it is a devourer of crabs and it lives voraciously on other bottom-dwelling forms. It is an expert in digging out flounders and no doubt it was hurrying for its dinner, for I could see no other cause for its unseemly speed.

The drifters were common and they had a special technique of their own. They always came with the tide and gave themselves fully to its movement. There was no struggle for them or frantic waving of fin or tail. They simply relaxed and allowed themselves to be carried with the current, stirring only enough to maintain their downstream direction and their position a fraction of an inch above the bottom. This was the mode of the Norfolk spots. Their silvery, slightly golden bodies were readily identified by the single black beauty spot located just above the pectoral fin. Their human counterparts were the Sunday strollers or, more exactly, the downtown window shoppers, for the spots often momentarily delayed their leisurely flights to inspect temporarily and lazily some portion of the bottom which, to my unpracticed eyes, looked like any other; and, like window shoppers, they never bought anything but soon gave themselves up to drifting again. At times their numbers were surprising, for on occasions they slowly streamed by in seemingly unlimited quantities. Peculiarly, I never saw them going out with the tide and it is possible they used other paths on their return.

Other species came and went in gangs and mobs. When they appeared it was in close-packed armies, body against body, rank behind rank. This was the way I first saw the croakers. I had never thought of them as operating in dense schools but during the time I spent in this one place I saw them thus several times. So crowded were their forms that when they wheeled or turned the sunlight caught their massed scales and the water for ten or twelve feet was momentarily caused to glow. Because of the haze I could see only those close by with clarity but a broad area of fitfully alternating light and shadow gave some hint of their extent. One school in particular must have numbered many thousands. They did not seem to mind my strange masked figure protruding above the sand, sending out a steady stream of glistening bubbles, but only veered slightly and passed on. Whenever I saw them I listened carefully, for croakers, as their name suggests, are able by virtue of a special modification of the air bladder, which acts as a resonating chamber, to emit a series of grunting sounds. I was hoping to observe some use of this unique physiological provision, but the fish were always silent and went quietly on their unknown errands. En masse the croakers presented an attractive sight; their individual bodies looked clean and trim and each fish was delicately colored with a slightly iridescent pearllike violet hue. They were predominantly silver but it was a silver tinged with delicate overtones. At times the violet verged on pink, then went through the spectrum to pale blue and back to violet again. The moment they are hauled out of water this play of light is gone and their bodies become metallic green and silver with no hint of the lovely shades revealed underwater.

It was only rarely that more than a single species of fish came by at a time though this may have frequently occurred just out of sight. There was none of the general mixing of types such as one sees universally in the tropics. When the croakers appeared they held the stage completely and the same seemed to be true for the other forms. There was one exception to this rule, however, and I have often wondered about the cause of it.

On the particular day involved I was resting in my usual favorite position waiting patiently for something to happen. For a long time nothing did, then slowly a few weakfish came leisurely up the underwater highway. The tide was nearly slack and they seemed in no hurry but were searching the bottom carefully as they came, presumably looking for food. Perhaps a dozen in all passed as far as I could judge from those which came close and by the more distant indistinct shadows of their companions. With them, but not together, were the swimming forms of several blue crabs; these were making good time and moved steadily along. One, startled by my coppery bubbling helmet, reversed its direction and went streaking out of sight. After the passing of the last crab there was an interval of perhaps five minutes when the water was completely blank and when the only movement was the slight shifting of the sand grains on the bottom.

Then, without warning, one of the weakfish—or at least its close brother—came hurrying out of the haze heading down the Bay toward Norfolk. It was followed in rapid succession by two more, by several croakers, a half dozen spots, and seven porgies. With them was the largest and only sheepshead I have ever seen. They were all swimming as fast as they could go and there was barely time to identify their shapes as they darted by. Whatever was the cause of their panic will never be known, for, although I watched for some time, nothing further appeared to disturb the abyssal calm.

One day I went out as usual to find my float and to learn more about the pilgrims on my underwater road. But when I arrived at what I thought was the location the buoy was nowhere to be seen; there had been no storm, so I assume it had been run down by a launch in the dark and destroyed. Although I made soundings and found the drop at the thirty-foot line without undue trouble I was never able to locate the exact place again. Other spots were, in comparison, barren of life. It was one of those secret and rare places fishermen eternally hope for but seldom find. As far as I know, the unmarked highway may still be there and its traffic undiminished. I suspect that many dozens of such favorite underwater routes may exist and that at certain times of the year when the great seasonal migrations are in progress these places must teem with life. If there were only some way to clear the opaque water they could then be plainly seen, and at the height of their use they would appear as long twisting living ropes extending from the green rivers and creeks to the depths of the outermost ocean. That this may not be pure conjecture is indicated by the millions of fish that suddenly appear and as mysteriously vanish each fall.

With the loss of the location of my submarine highway I had to be content with more infrequent findings, and in hope of greater variety moved the scene of my deep-sea explorations to Cape Henry, Virginia, at the very mouth of the Bay. There I had one of the most interesting days of my life. In a dark quiet depression where, for some unobvious reason, no strong tide seemed to reach, I found a unique spot. The depression was in relatively shallow water, only about twenty-five feet, and ran parallel to the shore a few hundred yards away. It was really a long narrow valley and was well protected from the ocean surge by its location below the general bottom level. Except for a gentle swing back and forth created by the ever-present ground swell, there was no difficulty in moving about and I could lie at ease on the ocean floor free of the constant danger of being carried away. It was an ideal place for intimate exploration.

I soon discovered that the obvious advantages of the location had long been anticipated by a host of beings other than myself. For, when I lowered my body to the soft sand floor and lay prone so that I could peer about, I found to my surprise that the bottom was carpeted with starfish. There were hundreds of them and there were several species. Wherever I crawled their symmetrical bodies were creeping over the sand or lying partially buried in the silt. I had never in my life seen so many in one place. It was a literal starfish heaven.

Starfish in the Chesapeake Bay are not too frequently found and they are restricted to the very salty region close to the capes. And even there they seem to be abundant only locally. In finding my submarine valley I had chanced on what seemed the center of the Bay's starfish population. There were three species. By far the largest and most common was a beautiful type of rich royal purple. I identified it later as one of the genus *Astropecten*, and their dark bodies were plainly visible against the yellow sand. Most were small, only three or four inches in total width, but a few were as large as my fully spread hand. Scattered between their bigger brothers were a number of babies; one tiny fellow was less than one-half inch from the tip of one diminutive arm to the other.

It is easy to become lyrical about starfish; their very shape causes them to be intriguing and there is hardly anyone who has visited the seashore who has not been somehow pleased at finding their beautifully radial bodies cast up on the beach. But other aspects of their unique character are much more fascinating.

If one were to be walking in a daisy field and suddenly saw one of the blossoms break away from its stem, climb off, turn upside down, and then go creeping away through the weeds on its inverted petals, a visit to the nearest psychiatrist would most certainly be in order or, at the very least, the time would have come to give up alcohol in all its many and enticing forms. Yet, if I may strain the reader's credulity and risk the not unnatural suspicion of requiring professional services myself, this is probably a likely simile of a starfish's early history. For starfish are closely related to the crinoids, a once-mighty host of ancient sea creatures which thickly carpeted the floors of the ancient seas. Their fossil bodies are found in abundance in the Paleozoic rocks and from these prehistoric stones more than two thousand forms have been described.

Today a small remnant of these archaic beings still persists unchanged in the nethermost depths of our modern oceans; there in the black chill where nothing ever alters and one century is just like another they grow in the abyssal mud and ooze exactly in the manner of their ancestors. Crinoids resemble nothing so much as a sort of misplaced deep-sea lily. They have roots, a stem, and a body which are very like the calyx and petals of a flower. But it is a weird kind of flower, for the petals writhe and squirm as though in some sort of vegetable agony. Crinoids are a deception, for they are not flowers at all but stalked carnivorous animals. Their bodies have the same basic symmetry and most of the structural peculiarities of starfish and, except that their arms are more feather-like, they are remarkably similar. Most interesting is that certain of the crinoids do exactly what our hypothetical daisy was supposed to do; they climb down from their stalks and go creeping about the sea floor.

Thus, while there is no absolute proof, it is not unlikely that the starfish are modified crinoids which have abandoned their roots and stems for a freer and more vagrant life. The paleontological evidence indicates that this may have been the case. And, in further proof, there exists today in the Mediterranean and also, so it is reported, in the deep waters off the coast of New England a feather star, *Antedon*, which grows as a sea lily until it is half an inch tall. Then the flower part—the body and its tentacles—drops off and spends the rest of its life crawling around like a true starfish.

Starfish are the very soul of patience and deliberation. The specimens that peopled the floor of my purple-spangled depression were traveling slowly about, creepingly molding their shapes to the contours of the bottom. They seemed to flow across the sand, moving as though they had all the time in the world, as indeed they had. I lifted one from the bottom and held it for a while on my outstretched palm. For a second or two it lay quiet and then began slithering over my flesh. I could feel its feet reaching out in successive waves and their multiple tiny contacts produced a queer sensation. It reminded me of the tingling feeling given by a centipede in its progress across the skin.

The comparison is apt, for starfish, in common with their near relatives, the sea urchins, are the centipedes of the sea. However, they are better equipped than their dry-land counterparts and many of them have more than a hundred feet on each arm. Further, where centipedes rely on perfectly ordinary, if multiplied, legs, the starfish have gone one better and devised a method of locomotion all their own. A starfish's foot is a sort of flexible suction pad, a rubberlike tube capable of considerable contraction and expansion. By means of special muscles and fluids pumped in and out of the jointless legs, the feet are extended until they touch some object; there special nerve cells decide what the foot shall do. If satisfied with its finding, the sensitive end will attach itself, pass the message along to the neighboring feet, and then, held by its terminal vacuum, contract and thus draw the starfish along.

Now, starfish have no brains; their nearest substitute is a circular series of nerves around the periphery of their bodies. These nerves are in turn connected to the tube feet by ganglia. In this way coordination is achieved and thus it is that the message of an individual foot is transmitted to its neighbors. So a starfish is really directed by its feet and not the feet by the starfish. It is a sort of tail-wagging-the-dog situation. A starfish's head is in its feet, so to speak.

The degree of this coordination, however, is sometimes remarkable. Particularly is this true of the way in which some of them secure their food. Interested in determining the cause of such a congregation of stars in the underwater valley I began crawling around seeking a plausible explanation. It was soon forthcoming, for the valley floor was sprinkled with the empty valves of numerous small mussels and other mollusks. These along with other creatures are favorite items of diet.

The mussels are extracted from their hard shells by a clever system in which the starfish's feet play a prominent part. When during the course of its foot-directed wanderings a star comes upon a live shell it folds its arms and body about its victim and seizes upon the halves of the shell with its hundreds of tube feet. The feet contract and then a long steady tug of war begins. The mussel, of course, tightly closes its shell and resists with all its strength. In the meantime the starfish ejects its stomach and wraps it around its intended dinner. The poor mollusk stands the strain as long as it can but sooner or later its single adductor muscle becomes tired and it reluctantly relaxes its valves and is devoured and digested by the enveloping abdomen. It loses the unequal struggle because the starfish, having dozens of feet and innumerable foot muscles with which to work, is able to maintain an unrelenting strain. While one series of feet are pulling, the neighbors are resting, preparing to send in relief when they have gathered new strength. These in turn work until the others have recuperated. It is a satisfactory and never-failing arrangement.

Starfish are unusual in other ways, too, and a discussion of their feet by no means exhausts their possibilities. Unlike most other animals, they have no heads and for that matter no lungs with which to breathe. Instead, the upper portion of the star is equipped with numerous small branchial papillae, small nodulelike projections which extract the oxygen from the water and then convey it by tubes to the various parts of the body. The papillae are always in danger of being destroyed by small carnivorous crustaceans and other animals or of becoming clogged by debris and dirt.

To prevent this fatal possibility, the stars have evolved a clever system. All over the surface of their bodies are minute projections; these are armed and equipped with tiny clusters of hard bristles or with diminutive pincers. The moment some foreign substance or moving animal alights on the starfish's epidermis the pincers begin opening and closing or the bristles begin brushing the intruder away. If the pincers seize some portion of the invader's anatomy they hold fast and others secure additional grips until the creature is helpless. Then it is passed slowly along from one pincer to another until the underside of the body is reached; then the hapless prisoner is transferred to the waiting feet and thence to the mouth on the underside of the body.

And now that I have written this story I must qualify it all and add that almost none of it applies to the other two species of the drowned valley. Such is the danger of generalization and one of the hazards of describing natural objects. For the other types were ophiurans or serpent stars, sometimes called brittle stars, and they follow a somewhat different anatomical pattern. There were not many and in all I found only seven individuals. They were quite small, and in contrast to their larger purple brothers were exceedingly delicate and fragile looking. The largest, measuring all of an inch and a half in full diameter, was an inconspicuous green and I would not have found it except for its comparatively rapid movement. Like all the serpent stars, it possessed a small pentagonal body with long worm-like segmented arms. Not being blessed with tube feet like its close relatives, the ordinary starfish, the animal was rowing itself along by sidewise undulating motions of its limbs. It was making excellent time. I picked it up and to my chagrin two of the five appendages immediately broke off and lay writhing in the sand. I let it go again knowing this was not the tragedy it seemed, for serpent stars, like all their family, have the power quickly to replace broken or damaged parts, a convenient ability.

The other species was one of the loveliest starfish I have ever seen and it is known by the rather lengthy technical title of *Ophiothrix angulata.* So seldom is it seen by land-dwelling humans that it has never acquired a common name. Its entire body, arms and all, would barely have covered the print occupied by its Latin designation. Attached to its greenish pentagonal central disk were five slender purplish arms covered with rows of feathery branchial spines. It appeared as nothing so much as some animated serpentine snowflake strangely endowed with life and transferred to a residence beneath the sea. It was so delicate and so fragile that I refrained from touching it lest by the awkwardness of my fingers it fall apart. The last I saw of it, it was daintily creeping its way between two small reddish finger sponges and thence out of sight.

I lingered in the submarine valley as long as I could. But soon the chill forced me to go to the surface and to the warm sunlight again. Intending to return, I left my equipment in a nearby shed and took the Baltimore boat home. Before I could return, a summer hurricane roared out of the Atlantic Ocean and ravished the coast from the Carolinas to New England. Tons of wave-driven sand poured over the sea bottom and permanently changed its contour. Though I dived time and again trying to find my star-studded glen, or another like it, it was forever gone. Somewhere, down beneath the bottom, the bodies of the purple stars are probably lying just as they were in life, overwhelmed by the drifting grains. Out in the dim reaches of the Bay close to the ocean there are likely to be dozens of similar valleys where starfish carpet the sand like constellations in the sky, but I have never found one. There, in the cool and the dark, they undoubtedly creep about, feeling their way on their extensible, sensitive feet and doing those things which starfish have done from time immemorial.

CHAPTER 16

THE WORLD OF THE WORMS

I HAVE NEVER BEEN PARTIAL to worms and, for all my inherent interest in natural history and the creeping crawling things of this earth, have seldom been able to become enthusiastic about their often peculiar and bizarre forms. This is a feeling shared by mankind in general and is probably deeply-rooted in the age-old but completely unwarranted distrust of anything which moves without benefit of legs, wings, or fins. Serpents also have unjustly been included in this category and have long been penalized in human regard for their extremely efficient but unorthodox mode of progression.

However, since going undersea, I have considerably

modified my opinion of worms and now accord them a full, if slightly reluctant, respect. I still do not pretend to know much about them and find that this ignorance is shared by the majority of my biological brethren. There are, of course, a number of excellent treatises on the anatomy and identification of worms and every college student has dissected their dreary and unlovely pickled bodies in the laboratory. But there is a singular silence in literature about their strange habits, about their appearance in their natural habitat, or about the slightly weird world in which they live.

To the average land dweller the word "worm" generally connotates a slimy, attenuated, reddish and featureless being which creeps beneath the grasses and roots of gardens where it conveniently performs some half-defined task of keeping the soil light and fertile, and which occasionally is useful for bait. The word also brings to mind funerals and sundry uncomfortable and mournful phrases of poetry and prose about the ultimate fate of man. Such dreary expressions as "the conquering worm," "our mortal bodies food for worms," and the like have, by their very unhappy implication, cast an unpleasant aura about the entire subject. This is unfortunate and puts the whole matter in an improper light.

I think it is about time that some few words are uttered in defense of worms and that some of the common misconceptions about their nature be dispelled. For, in spite of their reputation, worms are not necessarily slimy, nor are they usually smooth or featureless. Indeed, some are exceedingly beautiful and given to brilliant and delicate colors. Still others are very active and move about with far more energy than most animals; and not a few are highly intelligent creatures efficiently adapted for the portion of the world in which they live. Wherever there is sufficient moisture to maintain them, the worms have capably established themselves and often thrive in untold numbers.

It would not be an exaggeration to say that the Chesapeake contains some millions of individuals. They are found in all locations from the fresh waters to the salt, in the open Bay and in the smallest estuaries and creeks. Every acre of seaweed is alive with their creeping forms and the Bay bottom is pierced with their hiding places. The wharf pilings offer them haven and they cling to the hulls of ships. Every oyster bar has its quota of vermian inhabitants and the winding channels in the salt swamps and marshes are among their favorite haunts. Yet so secretive are their habits or so complete their disguises that they go almost unnoticed. Few Chesapeake residents know that their lovely Bay is literally crawling with worms. However, lest this give bathers and swimmers the shudders, it should be added that the Bay's worms do not care for human company and ask nothing more than to be left strictly alone. Unless one goes seeking them, they will never intrude or come where they are not wanted.

Yet, in proof of their abundance, I recall one evening just outside the snug little harbor at Solomons Island in Maryland. The entrance is partially blocked by a large seaweed-covered shoal which in ordinary times has only three or four feet of water over it. The yachts entering the harbor pass constantly a few yards either to right or to left. The bar is a favorite spot for blue crabs and for hosts of small minnows which hide in the weeds. But ordinarily except for those common forms of life there is no special feature to distinguish it from any other in the vicinity. Yet one summer night I rowed over its usually undisturbed surface and found it alive with swimming nereid worms. They were darting about in crazy circles, whirling their bodies round and round, going through the most violent contortions. The unusual activity attracted scores of large fish and they were gliding back and forth filling their bellies to bursting. I cannot estimate the number of annelids over this one bar but there must have been several thousand. Yet in the morning they were all gone, and though I have passed over this same shallow a hundred times I have never again seen even a hint of their existence. It is probable they were mating.

I became conscious that worms can be beautiful quite by accident. Far out in the lower Bay I once found a group of pound-net stakes which had been set up several years before and then abandoned. Their underwater surfaces had become covered with all sorts of marine growths and I thought they might repay examination. Sliding down the line, I hit the bottom and walked over to their vertical bases.

They were indeed interesting. Long festoons of lacy red seaweed hung gracefully from the wood and waved gently in the surge. The sun filtering down through the green caught the transparent tendrils and lighted the delicate tissue with a soft crimson glow. Between the fronds the fawn-colored anemones extended their filmy tentacles in hope of some small and unwary deep-sea victim. The feathery legs of the barnacles were rhythmically drawing in the water with its microorganisms; they appeared like small white volcanoes from which amber smoke momentarily issued only to be inexplicably withdrawn again. Between the barnacles and seaweed, clusters of bottle-green tunicates hung like bunches of unripe grapes and patches of lacelike calcium denoted the presence of colonial bryozoans. Small brown mottled *Panopeus* crabs crept between the attached organisms, gleaning little edible articles from the crevices and numbers of glassy prawn and other small crustaceans were hiding insectlike in each opening big enough to hold them. Every inch of space on the stake was taken up by some creature or plant, and so fierce was the competition for a place to live that some of the early inhabitants were all but buried beneath the bodies of later arrivals.

Such was the condition of a number of oysters and mussels crowded together in a massive cluster near the bottom. They were blanketed with a filmy growth of bright yellow arranged in small circular patterns. At first I could not make out just what the material was but when I dropped to my knees and brought the glass of the helmet close I saw what appeared to be a lovely garden of small translucent passion flowers. Each circular portion was a separate blossom and the individual petals were lacy and feathery. In the subdued light they glowed with a soft brilliancy. In the center of each was a group of long slender filaments resembling a cluster of orange-tinted stamens. The stems of the flowers, however, were not rooted but lay prone where they were seemingly glued to the shells that bore them. The blossoms were about the size of small daisies but, unlike any earthly flower, the petals were pulsing and quivering as though in deep pain. Against the dark-green water and the black shells the effect was striking.

While I watched, a small fish swam by and, attracted by some unseen morsel, darted close to the pole. As it passed, its shadow fell across the flower-sprinkled shells and in an instant every blossom was snatched back into its stem. In a twinkling the garden was gone and there was only a dark cluster of oysters covered with a multitude of dull-brown, parchmentlike tubes.

For a long time after the fish was gone there was no movement. Then, presently, a little spot of yellow began to show at the end of each cylinder. Cautiously, the feathery petals were thrust out only to be withdrawn again as the light was momentarily dimmed by a passing cloud. But gradually one blossom after the other slowly expanded and assumed again its flowerlike disguise.

The tubes, of course, were worm burrows and the blossoms and stamens, the animals' gill filaments and cirri. By means of these they extract oxygen from the water and determine the nature of the small micro-beings which they draw out of the Bay for food. Because of the ever-present danger of being eaten by fishes or destroyed by sharp claws of hungry crustaceans, each of the gill filaments is equipped with minute reddish eye spots so that the tender tissues may be warned and retracted out of harm's way at the least hint of danger. Imagine the effect of having a dry-land flower suddenly jerked out of one's hand before it can be plucked!

I gathered several of the shells with their attached worms and carried them to the surface. By gently pulling the tube apart I uncovered the entire animal—a short creature composed of about sixty greenish segments. From its anatomy I tentatively identified it as belonging to the genus *Sabella.*

Later on the same fish stake I found another flowered worm, *Hydroides,* a different type. Its gills were of a deep purple and one of its gill filaments was flattened to form a neat door which fitted exactly into the entrance of its chamber, much like the operculum of a snail or the trap door of a spider's nest.

The Chesapeake's worms, like human beings, may be classified by a variety of unscientific but descriptive systems. In the matter of pulchritude they belong to one of three categories: the beautiful, exemplified by *Sabella* and the purple-gilled *Hydroides;* the tolerable; and the downright repulsive—depending on one's degree of squeamishness and prejudice. Some can be two or even all three at once. The so-called opal worm found in the swamps and mud flats of the lower Bay is an example of this sort. The beast is an exceedingly slender animal, sometimes growing to fifteen or sixteen inches in length. Removed from its clinging mud and sand it squirms and writhes in a most unlovely manner, winding its body in and out of a serpentine spiral. But it is one of the most highly colored creatures I have ever seen. Waves of iridescent light play interchangeably over its threadlike body giving it the shimmering quality of the jewel that has given it its name. Why such exquisite colors should be wasted on a worm which spends all its hours in the wet black sand is one of the mysteries.

In contrast to the opal worms, which belong to the large and populous association of tunnelers and burrowers, are the smaller and more select group of swimmers and open-water wanderers. Not for them is a life of grubbing in the ooze and silt of the Bay bottom. Instead, they have given up all semblance of normal wormhood and have taken a free pelagic existence.

I first came upon them in the open waters between the capes. My boat had been becalmed for several hours and the waves had subsided into a slick oily ground swell. Quite a number of marine animals had come to the surface and were idling with fins above the surface or were lazily basking in the sun. Among them was a small school of menhaden which were excitedly swimming in a close circle. Their activity indicated some sort of marine food, so I launched the ship's dinghy and rowed over to see what it was. On my arrival the alewives scattered and at first I could see nothing in the green depths. But presently I discerned quantities of tiny needlelike motes milling about at surprising speed. They were almost transparent but the sunlight, catching on their glassy bodies, reflected brief sparks of light.

With a fine net I collected a number, brought them back to the boat, and put them under a lens. They were adult *Sagitta,* or arrow worms, strange little beings hardly a quarter of an inch in length, bustling about in their dish as though their lives depended on their speed. Their energy, like that of so many swimming worms, was all out of proportion to their size. Under the glass they resembled small fish rather than worms and the resemblance was heightened by their possession of two sets of paired horizontal posterior fins and a finny flat tail. The head, however, was typically worm-like and was ringed with long sharp sickle-shaped bristles surrounding the mouth; these are used in seizing their prey. For all their diminutive size the arrow worms have succeeded in establishing themselves in vast quantities over large areas of the ocean from the shallows to the greatest depths. This wide distribution has been aided by their agile swimming and the fact that each individual is complete in itself. They are able to propagate their race by self-fertilization; each arrow worm carries its own eggs and sperm, and the eggs are usually fertilized internally before being excreted for further development. In the wide spaces of the open sea where storms and currents may scatter such tiny creatures over immense areas this is a thoughtful provision to ensure the continuance of their kind.

Night is the time to see swimming worms. During the day they lie quiet in their dens waiting for the last golden rays of the sun to disappear and for the waters to turn gray-green and then inky black. Then they pour out of their burrows and go winding about on their nocturnal errands, seeking their food or doing whatever strange things prompt them to go coursing through the danger-filled Bay.

In the hope of observing them at first hand I once spent several hours in a large area of eelgrass sitting on the bottom of a little open glade just large enough to hold my form. As the dark settled, the gently waving fronds of the grass became indistinct and then blurred altogether. But presently as my eyes adjusted themselves to the gloom I could discern individual blades temporarily lighted by the phosphorescence as the gentle current brushed small animals against their edges. In the background glowing streaks and curving lines marked the courses of a multitude of small minnows as they dashed about after the small crustaceans that were beginning to swarm. I could feel their feathery bodies bumping into my bare arms and clinging momentarily to the hairs of my legs. Once, several small jellies drifted slowly by and then flared into greenish brilliance as they touched some object or were agitated by some unknown event important only to jellyfish.

Then out of the gloom between the black aisles of the shadowy, ghostly, half-lit eelgrass appeared a golden luminescence which, indistinct at first, became quickly clear. The glow resolved itself into a narrow band of phosphorescent light which curved sinuously back and forth in long undulating arcs from one side of the glade to the other. From the light it was creating I could see the maker plainly. It was a large nereid and the agitation of its hundreds of vibrating setae arranged in parallel rows on each side of its body caused it to blaze. So fast was the movement of the worm, it was trailing a streamer of fire for several inches behind its tail. The animal circled once, then fled in a long curve for the shelter of the fronds.

Presently it was joined by another and in the next half hour I saw six or seven of their weaving bodies. As the last four came by I quickly turned on my underwater lamp to identify them. They were all male clam worms easily distinguished by their general iridescent reddish-brown color and a vivid girdle of scarlet across their middles. Contrary to my expectations, the light did not seem to bother them and they swam on as if it did not exist. One halted temporarily at the base of a frond of weed and squirmed through the narrow roots as though seeking some sort of food. But in a moment it was swimming again, made a circle in front of the light, and then went hurtling off in the dark.

Once I had become used to their slightly weird and creepy presence I began to see a little beauty in their forms. Accentuated by the rays of the light, with the gracefully curving blades of the waving weed as a background they presented a surprisingly artistic picture. The contrast of their weaving, bright red bodies against the green, together with the curved vertical interspacing of deep-black shadows, was startling. None of their motions was unharmonious and they exhibited perfect control in all their actions. In the realm of line and color they were without noticeable fault.

The unexpected beauty of the *Nereids*, however, was

not shared by another nocturnal swimmer which I saw later on the Eastern Shore in some shallows near Cape Charles. I was not thinking of worms at all at the time but was interested in watching the antics of some courting pipefish which were cavorting near the surface of a clump of seaweed. The pipefish were holding my entire attention when I suddenly became aware that some unusual creature was intruding on the scene. From between two fronds of grass there issued what appeared to be a long, flesh-colored, flat ribbon about three-quarters of an inch in width. The ribbon was vibrating at the edges and seemed to flow through the water. Six inches appeared, lengthened to a foot, and then, to my horrified gaze, a whole three feet of quivering worm appeared. There was no visible head, only a tapered forward end with a narrow lengthwise slit where the mouth might have been. Happily, the light did not seem to interest it for it kept steadily on its course and in a second or two was out of sight in the dark.

I was so startled that it was some seconds before I could collect my thoughts and realize that what I had seen was only a specimen of the ribbon worm, *Cerebratulus,* which is found all up and down our Atlantic coast. It normally lives in the sand just below low-water mark but occasionally undertakes nocturnal swimming excursions for purposes which are obscure. *Cerebratulus* is one of the largest worms in the world and full-grown examples have been found nearly ten feet in length! Most of the Chesapeake Bay individuals are well under this size although it is entirely possible that in some favored localities monsters of this dimension may be found. I have dug small individuals from the sand in the lower portion of the Bay near the capes.

By far the majority of Chesapeake worms, not including the multitudinous parasitic worms of which there is no end of types or of the amazing places in which they live, belong to the tribe of diggers and burrowers. On the bars and tidal flats they maintain large colonies, veritable vermian villages, so to speak. Many have no permanent residence and, like ordinary earthworms, are forever prowling through the soft silt. Some literally eat their way along extracting their nourishment from the mud and ooze, digesting small particles and organisms of the soil. Others build well-constructed tubes and permanent homes in which they dwell all their lives. They are the prototypes of that sort of human being who centers his whole existence about a single house and becomes so adapted to his routine that to move elsewhere would be unthinkable. Or possibly they are in the position of the fellow who is persuaded by ambition or the prodding of his wife to invest in an exclusive neighborhood and then becomes, sometimes all unknowing, so attached to his mortgage and to his peculiar brand of suburbia that he could not and would not move if opportunity offered.

Most of these worm suburbias, like the human ones, are fringe developments. They are neither in the deep sea nor on the dry land but halfway between. Tidal flats are preferred or the area just below low water. Some exist without external evidence save, perhaps, a number of small holes in the sand. Others are plainly evident, the house of each resident well marked. In this last category belong the homes of *Diopatra,* the plumed worm. Almost every sand bar in the saltier lower Bay has dozens of them and in choice locations the colonies number thousands. The entrance of each house is marked by a turret of parchment and sand usually camouflaged with bits of shells and small sea wrack.

Plumed worms, like some other dwellers of outlying districts, are well dressed and, in their own way, rather handsomely adorned. Their bodies are reddish brown, stippled with gray spots, and they shimmer with opalescence. In addition, they sport a number of long red plumes and a variety of cirri, tentacles, and palps. Unfortunately, all this adornment is lost, for no one ever sees it. However, it is not altogether in vain, for the feathery plumes serve the useful purpose of gills and the cirri and tentacles aid in the important task of detecting and securing food.

Although *Diopatra* are good-sized worms, being at full maturity all of ten or twelve inches in length, they are not easy to see. Their burrows slope obliquely into the ground for three or four feet, and on the slightest disturbance they retreat to the farthest corner of their parchment-lined dens. Protected by the depth of soil and warned of the approach of potential enemies by the sensory tentacles and cirri, they lead a safe if unexciting life. They are satisfactorily equipped for just about any normal contingency. When the tide falls they doze away in their water-filled chambers until it rises again and brings with it the drifting hordes of small crustaceans, larval fishes, and other tiny beings which make up their diet. They have only to lie at the entrances of their tunnels until some tidbit comes within reach; and with the abundance of tidal life they seldom go hungry. They are in the enviable position of being able to live without working too hard and have only to reach out their figurative hands to receive.

One of the weirdest worms of the Chesapeake tidal flats

is *Polycirrus.* It lives just beneath the low-water line in the sand with its head and forward end protruding from the soil. The upper portion of its body is blood red and the head portion is equipped with long crimson, threadlike, tentacles which spread in all directions. The tentacles writhe like tortured serpents; they expand and contract as the blood surges through them. With undulating rhythm they sweep the sand and water in search of prey. In their small way they are among the earth's most grisly creatures. Only the fact that they are only two or three inches long saves them from being quite horrible. In large scale they would be fitting inhabitants of Mars or some other remote planet, or perfect subjects for the amazing-story type of magazine.

Yet their Gorgon-headed, medusae-serpentine tentacles are a perfect adaptation for their special needs. Fixed in the sand as they are for protection against larger creatures and thus unable to pursue their prey, they have equipped themselves with a satisfactory means of securing their suppers and gathering the necessities of life.

In a world where all normal terrestrial values are subject to strange contradictions and paradox; where animals play at being flowers, and flowers are carnivorous; where blossoms get off their stalks and walk around or blithely swim away, the sea worms add the final weird touch. There is almost no limit to the variety of their adaptations to solve their special and unusual problems or to the queer places in which they make their homes. In the matter of securing food they have utilized all the normal methods and have originated a few of their own besides. With horny jaws and sharp bristles they seize their prey; or their tentacles, like octopus arms, ensnare unwary beings. Some, such as the phosphorescent *Chaetopterus,* operate underwater vacuum pumps by which they forcibly suck their living from the sea; others gather small organisms by use of hundreds of minute waving hairs, or cilia; still others, including the enormously long ribbon worms, catch their food by the glue of their mucus, trapping it on the surface of their sticky proboscides, as insects are snared by sweet molasses on flypaper. Many worms ingest mud and literally eat their way through life and not a few have become suckers of blood and animal juices after the manner of the leeches.

There is almost no limit of specialization to which the marine worms have gone to achieve their peculiar position in life. By their diversity of habit and shape they have established themselves in places forbidding or impossible to more complex kinds of animals, and there is hardly a corner of their drowned world where they have not penetrated with success. On the Bay bottom in the clinging mud and gritty sand or in the vast reaches of open water and even in the flesh or on the bodies of other creatures, they have created their unique domain. Theirs is a strange and unreal existence in which all the regular rules seem subject to revocation, in which the abnormal is the normal, the bizarre the commonplace.

CHAPTER 17

COMPENSATION

ANY WRITER DEALING WITH compensation labors necessarily with two literary hazards —the difficulty of making a suitable beginning and the suspicion of imitating Ralph Waldo Emerson, who has written one of the perfect treatises on the subject in the English language.

Emerson, however, was primarily concerned with the good and bad in his fellow man; his composition was occasioned by indignation against the hell-fire and brimstone sermon of an exceedingly orthodox minister. The excuse for an attempt even to consider a similarly titled effort lies only in the fact that there is no interference with Emerson's excellent thesis and that the goodness or badness of man is, in the present instance, completely unimportant and irrelevant.

The problem in beginning, let alone ending, lies in the considerable choice of illustration. There is no aspect of the Chesapeake Bay, nor the life that teems in its waters or along its borders, which does not provide some demonstration, great or small, of the natural principle of compensation. Every individual micro-being, every bird and mammal, every fish and crustacean, the plants and seaweed, all are affected. One can start with any object and find in it some example on which to base a considerable philosophy; the compensatory law, like that of gravity, is omnipresent and inescapable. The beginning point may be anywhere.

Strangely, few books on natural history, or on geography, and certainly none on the Chesapeake Bay, have given the subject even the slightest consideration. I am certain not one person in ten million has ever regarded the Bay as a demonstration on a large scale of this principle, yet such consideration immediately endows every object, even the waters themselves, with a hitherto unsuspected significance.

It was only by sheer accident that I began giving the matter any thought at all. The sequence of events that are the key to this story started in an exceedingly simple way. I was sitting on the beach, alone, idly musing, not being concerned about any particular thing, merely enjoying the late afternoon and watching the long shadows steal across the sand. It was mid-June and the beach was deserted, save that back in the trees a night heron was circling in the fading light, waiting to begin its nocturnal activity. The day had been very calm and the darkening ripples that lisped on the beach were so slight as to be barely perceptible. The smooth water had reached its height and was slowly and gently slipping away again. As I watched, the uttermost line of water retreated, exposing a worn shell here, another there, then a line of sand, and soon a diminutive bar, and finally a small expanse of damp flat silt marked by the water currents and the trails of previously hidden sea creatures. At that point the darkness and night became complete.

The tide was falling and night had come; these were events which had occurred a hundred million times and would occur a hundred million times again. It was the least unusual of happenings, a fact so repeated as to be accepted without comment; the very assuredness of the event made it commonplace. Yet somehow this ordinary happening began a train of thought based on the obvious observation that if the tide would fall it would rise again, that the night was followed by the day, that for every action there was an equal and opposite reaction. I could not help but recall the importance my old physics teacher put upon this ancient and unassailable theorem, how for every down there was an up, for every in an out, how heat was the corollary of cold, how every positive produced its negative, how if the south attracts the north repels, how a gain in power is a loss in time, how a centrifugal force is followed inevitably by a centripetal result.

Two thousand years before the teacher was born and almost as long before his now glorified science was more than the plaything of a scattering of medieval alchemists, the Chinese had grasped the significance of the duality of nature and had expressed it in their philosophic concept of yin and yang, the so-termed male and female principle. The whole of nature is cause and effect; the simplest action begins a train of events which, like the pebble cast into a pond, sends out its ever-widening ripples. Thus, because of a Chesapeake tide, these words were written, the editor and the proofreader were put to much trouble, the printer and binder were made busy and you, the reader, are made aware that some several hundred days ago the water retreated and exposed a sand bar.

In other ways the ramifications of that single tide fall are still being expressed, if in constantly diminishing degree. For not long after, while I was still dreaming over the twofold and opposite nature of things, I idly picked up a cluster of ordinary oyster shells. They were seemingly no different from any of several hundred other clumps and I carelessly tossed them into a bag with a view to opening them later and having a few fresh oysters for dinner. Later, under the light, the shells proved a source of interesting speculation. Several contained, as expected, perfectly good oysters; another, and the largest, held instead the newly dead body of a small brown fish curled tightly about a cluster of glistening, damp eggs. The creature was a blenny, a small fish common all over the Bay, and surprisingly, a male; it had died from lack of water which had drained from the shell and allowed it to perish from dehydration.

It had prepared for every contingency. Inside its calcium fortress it was ready to maintain itself against all comers; safe from the jaws of hungry and larger fishes, guarded by a jealous parent, the eggs would have matured and a host of young blennies would have gone their way.

All would have been well except for the one impossible happening which no natural instinct could have anticipated—that the tide should have fallen and that I, the writer of these lines, should have been there at that exact time. And so, because this tide retreated, the whole effort of a blenny courtship was wasted; the eggs would never hatch; all the numerous future progeny would never see the light of day or know the pleasure of swimming in green waters; hundreds upon hundreds of small crustaceans would never be devoured to provide sustenance for the blenny's young and would live longer lives until their energy would be used in other directions.

The interesting fact that the blenny was a male was a demonstration in its own small and peculiar way of the duality of nature; the blenny itself was only half of a small entity; its guarding of the eggs but an equal share of a total burden, the ensuring of its kind. The probability that the assumption of that burden was only blind instinct does not make it wholly different in purpose, nor are its cause and effect much unlike either in substance or in quantity the phenomena of the love of a man for a woman with all its connotations of essential equality and of sharing. The occurrence of an unthinking fish in an oyster shell protecting with its life the eggs laid by its mate is no different in substance from the sacrifices of human parents that their young may prosper. Thus it is that the dual aspect of all living beings is reflected in every particle; great and small, complex or simple, each bears in some portion some expression of the entire scheme of things. Each individual is a unit of total pattern, and all subscribe, consciously or not, to this elemental truth.

There are no favorites; for blennies, or oysters, or men, every gift is accompanied by a compensating defect. The blenny lost its life because it depended on an oyster shell; and an oyster, in its turn, had given up intelligence and mobility that it might lie in safety in a prison of lime. Contemporary man is cheating himself of future oyster dinners by his very efficiency in collecting them; he cannot take and not give; success in one day means failure in another. There are no monopolies or exceptions, and the price of oysters on any restaurant menu will testify to this reality.

It is an intriguing thought that every thing in nature carries within itself a full, if small, reflection of the sum of nature. We are all of one essence; and, except for size and complexity, there is little difference between a flea and an elephant, and it is unlikely that the flea is less efficient for being small. A drop of dew is round for the same reason that a planet is; and a lichen on an arctic rock is no less alive than the manager of an atomic industry. The child of a Park Avenue matron has the same claim on eternity as the young of a Susquehanna shad, and the fact that the shad fry has a million brothers and sisters and the matron's boy but one is an indication only that both are governed by the same inflexible laws which decree that the numbers of progeny for a given species are in exact proportion to their chance for survival.

Thus the world, or the Chesapeake Bay, or the smallest jellyfish in its waters may be viewed as an expression of a universal equation, a mathematical proposition which, for all the seeming exceptions, is nevertheless absolute. There is always a balance, although the balance, like the Roman letter *x*, may be a cleverly concealed quantity. The actions and reactions of each being are but a recitation in whole or in part of the fundamental proposal; every relationship is a correlative of every other.

It is only in the lesser events that some understandable simplification of the complete formula may be grasped; the tide falling on a sandy beach, a dead fish in an oyster shell. Yet these minor happenings, by their very character, are more vivid; just as two times two is more expressive and is as equal a verity as a thousand-page exposition of Mr. Einstein's concept of relativity.

That this is so was indicated the same evening after supper when, after having tempered philosophy with a physical fortification of the identical oysters that prompted this discussion, I returned to the same beach and sat down on the sand to see what would come next.

The tide was still out and it was very dark. Overhead the cloudy luminescence of the Milky Way softly emblazoned its path across the zenith, and to the east some unknown but brilliant star shed its lambent flame over the water and left a delicate and wavering tracery across the quiet black depths. There was no wind and back in the woods a monotone of nocturnal insects droned steadily along; occasionally a short, liquid, and softly repeated note revealed the presence of some half-slumbering bird. For a long, long time there was no other sound.

Then somewhere, off in the distance, there came a vibration all out of keeping with the whisperings of the night. The tremor was more felt than heard, and for a brief second I imagined that it was followed by a hush, a temporary silence as of anticipation, like the quiet that suddenly falls upon a concert hall an instant before the first note.

Then it came. Far up the shore, hidden in the dark, something had happened. There was a dull booming, followed by a muted rumbling, the sibilant sound of disturbed liquid. Though I could not see it, I knew what it was. Some tree, poised precariously on the eroded, wave-eaten bank had finally given way and fallen prone into the water.

I found it in the morning, a large oak, still alive and green. It lay on its side in the salty Bay, its branches crushed and broken. For seventy or eighty years it had stood straight and tall, flourishing through the years, spreading its branches, blossoming in the spring and shedding its seed and its leaves in the fall. Storms had beat against its trunk; rain and hail had lashed its limbs; the sun had poured upon its leaves, which by the alchemy of their chlorophyll had become converted to twig and branch and bark and stem. From soil and air and light it had borrowed its being, accepting each season a loan on its life, an annual increase in principal, the only interest a payment of foliage in the fall.

For almost a century it had deferred its debt; through gale and squall it had stood; neither disease nor blight had marred its shape; nor had vicissitude pierced its shaggy armor. Its nemesis was the tide, the soft slow creeping tide that took away the last particle of crumbling earth, the last fragment of supporting soil, the final grain of sand. With not a quiver of wind in the leaves, the denuded roots had reached the ultimate quota of bearable strain; their fibers, the extreme and unsupportable load. Somewhere deep in the soil a rootlet had snapped, released its hold, and in sudden succession the whole subterranean system had torn asunder.

It was a simple event. The tide had turned and a tree had fallen. Yet think of the connotations of this unadorned statement. A flourishing plant had reached the end of its usefulness; in its full maturity it had accumulated the full total of its allotted time. It would soon become a sodden log, the bright green leaves bits of frayed brown tissue, the bark loose scaly gray powder. This is the manifest evidence.

However, it is not so obvious as that. In nature nothing is given; all things are sold or lent; for every gift there must be a return, for every debt a payment. Precisely as the flood follows the ebb, the warmth the cold, just as spring follows the fall, so failure accompanies success. A tree lives; it dies. It surrenders its leaves and twigs and branches to the soil that nourished it; its limbs return to earth and become soft brown loam and a home for burrowing grubs and hard-winged beetles.

But there are two sides to every coin and the God of Failure is Janus. The tree is forever gone; its shape altered and lost. But because the tree fell, because its leaves drifted to earth or were swirled away in the waters, because its limbs lay broken on the beach, its bark on the Bay floor, other life became possible. Existence is like a relay race in which the exhausted runner hands the baton to the next in line; the wand is the symbol of continuance. Today's blossoms are yesterday's dried leaves and the hum of tomorrow's bees is the transmuted nectar of the same flowers. Nor is it a literary stratagem to note that the song of some courting bird may be the distilled essence of some lowly grub which would never have lived if a tree, no different from the tree on the beach, had not fallen and become a moldering log.

That this is not pure rhetorical fancy but a considerable approximation of the truth is shown by the history of this same oak. On the morning after its sudden disaster there was no hint of its potential role. Instead, it was in the situation of a foundering ship, a mangled wreck of branch and limb, with vines for tangled rigging; and like the ship it was being deserted by its crew—by all those beings which had made it their home. Down the trunk was a scattered line of reddish ants marching in single file carrying their eggs to safety. Somewhere out in the limbs their once-snug aerial home had become soaked with salty water. And in panic lest their young should die they were struggling the long yards to shore over what to them must have been a tortuous and hazardous journey. In a sense they, too, were paying a debt, discharging a contract assumed unknowingly with the first breath of life. They were bound to the obligation that come what may—be it storm, enemies, or the almost unthinkable disaster of falling into the salty green Chesapeake—their race should go on, that the tribe or horde should not diminish by one individual. Thus, though it would have been simple to have climbed unburdened down the long way to individual safety, the welfare of the group came first. And it is likely, in compensation for this undivided and assiduous devotion, that this peculiar type of ant will persist unchanged when larger and less organized creatures will perish. As was pointed out earlier in these pages, a being is not less perfect for being small.

The fall of the tree was disaster for other beings, too. Several bright green katydids, clutching disoriented limbs, were surveying their disturbed world with curiously waving antennae. One of these, not satisfied with its situation, or perhaps disturbed by my presence, spread its wings and with the weak flight of its kind flew inexplicably into the wide-open spaces of the Bay. For a long time I watched it go and then saw it settle uncertainly on the waves. It never rose again, and there it probably drowned. It also was forfeiting its claim, was paying a penalty for being so specialized for an arboreal and earthy existence that once removed from its accustomed woods and meadows it could not distinguish green water from waving grass or quivering forest leaves.

Whatever birds had been resting in the limbs had long since fled, but tucked crazily on one limb was the inverted cup of a bedraggled nest. It was new and whatever eggs it might have contained were gone, thrown into the water and there destroyed. At first there was no sign of the owner but then late in the afternoon the continued presence of a gray feathered form, chirping petulantly in the broken branches, identified the type; it was a vireo, but before long it too disappeared and the tree was left empty and deserted.

For many months I did not see the tree again, but when at last I viewed its form it had undergone considerable alteration. The leaves were all missing and the bark had shaled away in large slabs exposing whitened and sun-bleached wood. It had sunk low in the water and the trunk was half buried in the sand. The waves slapped steadily against the slippery bole and some of the limbs were soft with rot. Most of the small twigs were nowhere in sight; they had, no doubt, floated away on the tide or been covered by the sand.

But the tree was no longer alone, nor was it deserted. In its death it was supporting a more varied life than at any time in its youth. The once-firm wood, so resistant to every invasion, was riddled with holes and with the burrows of hundreds of wood-dwelling creatures. Little by little they were gnawing its substance, drilling long tunnels and secret passages into the heart of their host.

With a thin stiff knife I began stripping away the bark and carving into the softened tissue. From dozens of hiding places I retrieved the diggers and the authors of these labyrinthine runways. They were an interesting collection and there was a variety of species from an assortment of orders. But for all their diversified structure they were alike in one feature—all their lives and their entire existence was predicated on the assumption that a tree would fall—in the woods, on a beach, or in the heart of a dark swamp. On this assurance these varied beings had undergone an extensive specialization, anatomical changes to meet this exact and as yet unfailing probability. And for thousands of years this certainty had never failed, for their fossil forms show that some have not altered for thousands of centuries; the trees changed; whole forests of species have become extinct or by slow evolution have gone forward to other types. But certain of the wood borers not at all. So satisfactory is their mode that they have retained it intact.

Here, too, is compensation of a kind. For the same exacting Nature which by her anatomical limitations forever binds a creature to an inflexible role nearly always ensures that the part can be fulfilled. While she is at times cruel and ruthless and seemingly unconcerned with the individual, every provision is made for the consummation of the total pattern. While it is useful, nothing is cast aside or heedlessly destroyed. Trees are bound to decay and wood borers to hasten that condition. An untouched and fallen tree is only a delay in a vital process; until the bark and fibers are transmuted and transformed there is no life again. Dead wood is dead wood only; but that same wood churned and chewed by insects, digested by worms, or dissolved by the rains is the substance of tomorrow's anemones and next year's ferns.

As I pondered over these mysteries and sorted the products of my varied collection I began to marvel at their exceedingly intricate adaptations. There were ants, not wholly unlike the ones which first fled their fallen tree, but larger and with bigger jaws. These jaws were strong and tough and equipped with razor-sharp chisels. They had commandeered the whole stump end of the trunk and had there carved a veritable city of tunnels and caverns. And there were scores of flattened light-brown roaches, slim darting shapes which scurried shadowily about and had their domain in confined crevices. They deserve our considerable respect, for they are among the most ancient of living fossils. As far back as the Paleozoic age, before there were any trees at all, they were propagating their kind and squeezing, as they do now, into the narrowest of narrow spaces. For all their unhappy and undeserved reputation they are among the earth's most successful inhabitants.

But it was the beetles that held the stage and had the leading parts in their roles of destruction. All over the trunk they were working, industriously sawing, grinding, drilling, abrading, and chiseling away their abode. In a way they were in the paradoxical position of the doctor striving steadily to cure his patients and put himself forever out of business. With every hole pierced, every fragment of cellulose chewed, they were hastening the time of their necessary departure. But, like the doctor who persists confident of a continuous supply of new cases, so these beetles merrily worked to grind up their tree and reduce it to fine powder.

There was an assortment of means to the common end; and most delicate and artistic were the engravers. These were small cylindrical fellows with fine abrading mandibles. Their special domain was the narrow space between the bark and the wood, and here they carved intricate patterns of winding and interconnecting tunnels. Their work created a lacy pattern of finely curved lines reminiscent of the designs on old and antique Moorish buildings or, better yet, the lovely etching seen so often on medieval and sickle-shaped scimitars.

Others were borers and with their chitinous chisels they drilled deep holes as clean and as exact as if done with fine steel augers. Then there were the honeycomb makers which, like those unimaginative and satisfied people who live perpetually in one community, concentrated their whole activity in a small area, working it back and forth, searching into every possible crevice until the entire section was a mass of interlocking chambers. So thoroughly is their work done, that the once-stout wood may be crushed between the fingers.

But whatever their mode or whatever their special form of carpentry, they have made, almost without exception, distinctive concessions to their kind of life. Big or little, long or short, they were all devoid of bright colors and were all exceedingly smooth. Except for some which possessed fine lines running the lengths of their bodies, they were virtual animate mirrors and in many cases were so highly burnished as to glisten like polished and well cared-for furniture. Also, their external parts were capable of being cleverly concealed and were so designed as to be readily tucked away and to fit their bodies integrally much as the disappearing wheels and undercarriages of high-speed airplanes. Vanishing legs were common long before bombers were thought of.

The simile is not altogether inept, for the fundamental purpose is the same. The sleek tapered bodies of supersonic rockets and the glossy smooth torsos of burrowing beetles are a concession, both, to the fact of resistance.

That one moves at lightning speed through the air and the other slowly through tough lignous wood is immaterial. It is as impractical for a wood-burrowing beetle to have bumps as for a jet plane to carry a turret or an observation platform. And so if one would fly through space with the speed of sound or move but an inch an hour in the depths of a decaying log, one must make a compromise, a compensatory contribution, so to speak, to the same essential laws; must forfeit variety for a basic standardization, freedom of form for efficiency.

Within the limits of the factors imposed upon them by necessity, the burrowers of the trees were nevertheless amazingly diverse, nor were they restricted to that portion of the wood above the water. At some indeterminate point about half through the trunk the terrestrial tunnelers acceded place to the marine drillers. For a little distance there was a sort of no man's land—a half-soaked region which belonged to neither; but beyond that point the sea forms took over completely and an examination of the tree showed that they were doing a more thorough and rapid job than their dry-land competitors.

Most important of these was an animal which is a living demonstration of the remarkable relationships that can occur, and which is a unique example of the lengths to which Nature will go to ensure that her processes continue unabated and undelayed. The creature is wholly an organism of the sea, yet its very existence is based on the fact that trees will live and that their remains will fall in or be carried to the alien salt water.

The beast is resident all over the world and there is no shore tinged by salt water where its activities are not plainly and sometimes painfully evident. In floating logs and in the timbers of ships it has traveled from ocean to ocean; many is the sunken hulk that owes its wrecking to this beast's peculiar proclivities. Like so many other beings which succeed in specialized niches, this animal pretends to be what it is not and its very name is a deception. I refer to the common shipworm or, as it is more properly termed, the teredo. The teredo is not a worm, nor is it particularly devoted to ships. Teredos were gnawing wood some millions of years before either man or his vessels came into being and will probably be doing the same when navies are but a memory. The passing of the wooden vessel has not affected the life of the teredo by one iota, nor diminished its numbers by enough to count.

The masquerade as a worm is, instead, only a deference to the same necessity which smoothes the anatomy of burrowing insects. The teredo is really a mollusk, a relative of the oysters and the clams, and its shape is the solution of the identical problem; one might say it is a streamlined oyster. Like the oyster, it has a shell though a much distorted and misshapen one. Also, like its close cousins, it has a delectable oyster taste; and, if you doubt my word, try one sometime. One's first teredo is no more difficult than one's first oyster and except for the fact that the teredo is long and thin and the other short and fat, there is little to choose between them. Both oyster and teredo derive their food from the same source, eating minute organisms drawn from the surrounding sea. The only difference is that one lives in a closed shell on a bar, the other in a calcium-lined tunnel in a piece of dead tree. Not one American in ten thousand, however, will attempt a teredo because of an unexplained and hidebound prejudice against any being which is long and narrow. Thus a potential delicacy is neglected; if teredo on toast were known and established, it would be an item for epicures.

Only snail-eating Frenchmen and Italians are possible prospects for this projected tasty dish—Frenchmen particularly, for any people which will cook and devour small sparrows, head, insides and all as I once saw done in a French village, can have no culinary inhibitions.

Teredos are very efficient in their carving techniques and there is almost no kind of timber they cannot riddle efficiently. They enter the wood when they are very small and increase in size as they continue; the location of their doors is often quite invisible. At their posterior end they possess a pair of rasping valves which enlarge the creature's chamber by first working across the grain and later boring along with it, a surprisingly efficient system. Thus a shipworm progresses backward into its home while drawing water, oxygen, and food through calcareous structures known as pallets at the entrance. The pallets safely seal the opening against all intrusion or enemies. By these means and the certainty of a never-failing supply of trees they persist happily through the years, unenvied, unassuming, and unafraid. Their compensation for living in the darkened interior of water-soaked wood is the freedom from fierce competition and the ease of being undisturbed and unmolested; they have traded danger for secure monotony, a peculiarly modern, if socialistic, human philosophy!

The underwater inhabitants of the fallen tree included, in addition to the mollusks as represented by the ubiquitous and successful teredos, specimens of most of the other important vertebrate orders. Beginning from the bottom of the social scale and working up the underwater society ladder, the tree, like the streets and houses of a large city, harbored an inextricably mixed population, the high and the low, the simple and the complex, all in one heterogeneous assemblage. Invisible to the unaided eye, but present in countless hundreds, living in every minute crevice and seam, and even on the bodies of the other beings, were the protozoans, the micro-animals, the metaphoric fleas and lice of the larger inhabitants.

The sponges were there, too, not as the familiar soft porous balls of our baths but as dull-brown and red or green incrustations. With them were the small bright yellow tufts of the *Cliona,* the boring sulphur sponges. Ordinarily they prefer the calcareous shells of mollusks but will drill their way into almost any solid material. Unlike the other borers of the dead tree colony they rely on a system of their own for their drilling. Some unknown chemical or solvent is used in place of jaws or rasps; here again is a striking example of the compensatory principle; *Cliona* have no organs or specialized tools for their task but whatever benign deity watches over them has provided an equal and as efficient a substitute.

The *Coelenterata,* that is to say, the jellyfish (or at least their immature forms, the hydroids), and the anemones were present by the dozens. Their small plantlike bodies were spotted all over the submerged wood and their translucent flowerlike tentacles were gently waving in the currents. Like the false flowers they seemed to be, they imparted the perfidious illusion that the trunk was coming to life again and was sprouting some strange sort of underwater blossoms.

The soft wood lying against the sand was the home of the next in the social strata. This was reserved exclusively for several types of worms; their precise identification, however, eludes me, for I am always appalled by their unlovely appearance and the complexity of their identifiable anatomy. There was no restriction, however, in locale, for the legions of crustaceans. They had taken up their abode in every conceivable crack and cranny. One fragment which I removed from the tree and laid on the beach, for hours gave up minute amphipods, copepods, and other forms as their dwelling place dried and they crept forth to seek their accustomed water.

Of all the major phyla only the echinoderms, the starfish and sea urchins, were missing and no doubt if the oak had fallen farther down the Bay in more salty water they would have been present also. Even the vertebrates were there, for while digging close to the bottom I disturbed a small goby, a plain little dusky-green fish whose adult size is hardly an inch in length. It was quite naked, after the manner of its kind, and did not have a scale on its entire body. It was equipped with a sucking pad formed from its ventral fins and with this it clung determinedly to the trunk. The tree apparently provided it with a good and easy living, for from its diminutive stomach I later extracted the complete bodies of several *Gammarus* (small amphipod crustaceans), an annelid worm, and a baby fishlet of unknown parentage.

Thus, out of one fallen tree was formed nearly a complete cosmos of living things; almost every class of animate existence in one degree or another made up its structure or contributed in some way to the organization or survival of its members. Directly or indirectly, the tree had become sustenance or the source of maintenance of a host of minor beings which in their turn provided life and form to still dozens of others. It was a sort of biological "house that Jack built" and if the sequence that began with the falling of a tree could be traced to its ultimate conclusion, there would be evolved a miniature pattern of the whole of life.

For there is a "life out of death" which has no relation to the obscure metaphysical explanation ordinarily given it. It is a nontheological phenomenon open for all to see and understandable, in part at least, by the simplest intellect. Neither the largest tree, the smallest flower, the lowest worm, nor the mightiest whale in the sea perishes in vain. To the least of these and to the greatest is given a universal, if not immediate, immortality. The life stuff that is the mysterious and unknown substance of all animate existence is never abated or appreciably diminished; for every loss there is a compensatory gain; and, as Emerson so beautifully expressed it, "God reappears with all His parts in every moss and cobweb."